High Entropy Materials: Challenges and Prospects

High Entropy Materials: Challenges and Prospects

Editors

Weidong Li
Peter K. Liaw

MDPI • Basel • Beijing • Wuhan • Barcelona • Belgrade • Manchester • Tokyo • Cluj • Tianjin

Editors
Weidong Li
University of Tennessee
USA

Peter K. Liaw
University of Tennessee
USA

Editorial Office
MDPI
St. Alban-Anlage 66
4052 Basel, Switzerland

This is a reprint of articles from the Special Issue published online in the open access journal *Metals* (ISSN 2075-4701) (available at: https://www.mdpi.com/journal/metals/special_issues/high_entropy_alloys_challenges_prospects).

For citation purposes, cite each article independently as indicated on the article page online and as indicated below:

LastName, A.A.; LastName, B.B.; LastName, C.C. Article Title. *Journal Name* **Year**, *Volume Number*, Page Range.

ISBN 978-3-0365-2406-1 (Hbk)
ISBN 978-3-0365-2407-8 (PDF)

Contents

About the Editors

Weidong Li obtained his B.S. in Materials Science and Engineering from China University of Geoscience (Beijing) in 2007, his M.S. in Materials Processing Engineering from University of Science and Technology Beijing in 2010, and his Ph.D. in Materials Science and Engineering from University of Tennessee in 2013. He has served as an adjunct assistant professor in the Department of Materials Science and Engineering at the University of Tennessee since 2018. Following graduation, he worked in R&D units of the ceramic, rubber and tire, and aerospace industries on a variety of things. His research interests generally lie in alloy design, integrated computational materials engineering (ICME), fracture and fatigue, and the mechanical behaviors of materials, specifically materials such as high-entropy alloys, superalloys, steel, and metallic glass.

Peter K. Liaw graduated from the Chiayi High School, obtained his B.S. in Physics from the National Tsing Hua University, Taiwan, and his Ph.D. in Materials Science and Engineering from Northwestern University, USA, in 1980. After working at the Westinghouse Research and Development (R&D) Center for thirteen years, he joined the faculty and became an Endowed Ivan Racheff Chair of Excellence in the Department of Materials Science and Engineering at The University of Tennessee (UT), Knoxville, since March 1993. He has worked in the areas of fatigue, fracture, nondestructive evaluation, and life-prediction methodologies of structural alloys and composites. Since joining UT, his research interests have included the mechanical behavior, nondestructive evaluation, biomaterials, and processing of high-temperature alloys and ceramic–matrix composites and coatings, with the kind help of his colleagues at UT and the nearby Oak Ridge National Laboratory. He has published over one thousand journal papers, edited twenty books, and presented numerous plenary, keynote, and invited talks at various national and international conferences. He was awarded the Royal E. Cabell Fellowship at Northwestern University. He is a recipient of numerous "Outstanding Performance" awards from the Westinghouse R&D Center. He was the Chairman of The Minerals, Metals and Materials Society (TMS) "Mechanical Metallurgy" Committee, and the Chairman of the American Society for Metals (ASM) "Flow and Fracture" Committee. He has been the Chairman and Member of the TMS Award Committee on "Application to Practice, Educator, and Leadership Awards." He is a fellow of ASM and TMS. He has been given the Outstanding Teacher Award, the Moses E. and Mayme Brooks Distinguished Professor Award, the Engineering Research Fellow Award, the National Alumni Association Distinguished Service Professor Award, the L. R. Hesler Award, and the John Fisher Professorship at UT, and the TMS Distinguished Service Award. He has been the Director of the National Science Foundation (NSF) Integrative Graduate Education and Research Training (IGERT) Program, the Director of the NSF International Materials Institutes (IMI) Program, and the Director of the NSF Major Research Instrumentation (MRI) Program at UT. Several of his graduate students have been given awards for their research and presentations at various professional societies and conferences. Moreover, his students are teaching and conducting research at universities, industries, and government laboratories.

Preface to "High Entropy Materials: Challenges and Prospects"

High-entropy materials emerged in the first decade of the 21st century, attracted the attention of metallurgists, and became a hot topic in the material science community in the following decade. The new fever began with high-entropy alloys, a family of multi-principal element alloys with at least five elements in equal or near-equal molar fractions. This family of alloys interested metallurgists soon after their invention, because the complexity in chemistry results in a disordered solid solution phase in certain instances, rather than multiple intermetallic phases, as expected from historically established metallurgical experiences. The effect can loosely be attributed to the high configurational entropy caused by the mixing of multiple elemental species in close ratios, which tends to minimize the Gibbs free energy to stabilize a solid solution. It is an exciting field to work in, as a multitude of compositions, microstructures, and properties have been reported and are continuously revealed. It is also exciting because the material space and category is expanded beyond those in which metallurgies have existed for hundreds of years. It is anticipated that plenty of novel alloys with excellent properties and real-life applications will evolve in this field.

High-entropy materials emerged in the first decade of the 21st century, attracted the attention of metallurgists, and became a hot topic in the material science community in the following decade. The new fever began with high-entropy alloys, a family of multi-principal element alloys with at least five elements in equal or near-equal molar fractions. This family of alloys interested metallurgists soon after their invention, because the complexity in chemistry results in a disordered solid solution phase in certain instances, rather than multiple intermetallic phases, as expected from historically established metallurgical experiences. The effect can loosely be attributed to the high configurational entropy caused by the mixing of multiple elemental species in close ratios, which tends to minimize the Gibbs free energy to stabilize a solid solution. It is an exciting field to work in, as a multitude of compositions, microstructures, and properties have been reported and are continuously revealed. It is also exciting because the material space and category is expanded beyond those in which metallurgies have existed for hundreds of years. It is anticipated that plenty of novel alloys with excellent properties and real-life applications will evolve in this field.

This Special Issue aims to provide a platform for researchers in the broad field of high-entropy materials to communicate their views and recent research on the subject. Reprinting it as a book is helpful for further raising awareness of the field and promoting communication among academic and industrial researchers.

Weidong Li, Peter K. Liaw
Editors

metals

Editorial

High Entropy Materials: Challenges and Prospects

Peter K. Liaw * and Weidong Li *

Department of Materials Science and Engineering, University of Tennessee, Knoxville, TN 37996-2100, USA
* Correspondence: pliaw@utk.edu (P.K.L.); wli20@utk.edu (W.L.); Tel.: +1-865-974-6356 (P.K.L.)

1. Introduction and Scope

Entropy is an important concept in thermodynamics, measuring the disorder in a system or, more precisely, the number of possible microscopic configurations of individual atoms or molecules of a system, i.e., microstates. High-entropy materials (HEMs), in terms of entropy, are a class of materials with a higher-than-usual degree of disorder in their microstructures. The high degree of microstructural disorder in these materials implies that they possess simpler microstructures than their low- and medium-entropy counterparts. A representative example is the formation of a single disordered solid solution in a multi-component alloy system, rather than several phases expected from the Gibbs phase rule.

The high-entropy concept was first proposed in the field of metals in 2004 by Yeh et al. [1], and the relevant materials would later become widely known as high-entropy alloys (HEAs). What interests researchers about HEAs is that the mixture of five or more elements in equal or near-equal atomic proportions could result in a single solid solution phase, which is unexpected given the established metallurgical wisdom. The mix of multiple elements in equal or near-equal atomic proportions also overturns the long-established alloy-making practice of utilizing one principal element and adding others as minor ones for property tuning, thereby creating a drastically huge compositional space for material researchers to work with. The research into the multiple different aspects of HEAs continuously reveals unique microstructures and promising properties [2,3]. All these facts are making HEAs one of the most promising fields of research in the material community.

The high-entropy concept is not limited to the field of metals; its scope includes other materials. High-entropy ceramics, high-entropy polymers, and high-entropy composites are emerging to enrich the HEM family. Predictably, prospects and challenges go hand-in-hand in each class of HEMs. This Special Issue has been launched to provide a platform for researchers in this field to report on the prospects and challenges they face in advancing HEM research.

check for updates

Citation: Liaw, P.K.; Li, W. High Entropy Materials: Challenges and Prospects. *Metals* **2021**, *11*, 1643. https://doi.org/10.3390/met11101643

Received: 11 October 2021
Accepted: 14 October 2021
Published: 15 October 2021

2. Contributions

This Special Issue is composed of eight articles, which have been contributed from various branches of HEMs and related fields, including five articles on HEAs, one on high-entropy ceramics and composites, and one on bulk metallic glasses.

In the five articles on HEAs, two focus on refractory HEAs, one designing a single-phase body-centered cubic (bcc) in the Mo-Nb-V-W-Ti system with classical empirical rules and studying their microstructures and mechanical properties accordingly [4], the other exploring the laser beam welding of the $Ti_{1.89}NbCrV_{0.56}$ and reporting on the influence of differently configured welds on the microstructure evolution and mechanical properties of the alloy [5]. In addition, there is one article attempting to tailor the composition in the $(Fe_{0.3}Co_{0.5}Ni_{0.2})_{100-x}(Al_{1/3}Si_{2/3})_x$ system to design cost-effective soft magnetic alloys with satisfactory mechanical properties [6], one article statistically analyzing the prevalence of intermetallic (IM) phases of 142 IM-containing HEAs and inspecting the inheriting relation of the IM phases in HEAs with what were reported in their binary and tertiary subsystems [7], and one article reviewing the potential of HEAs as biomaterials [8].

The article on high-entropy ceramics reports on the steps of making (Hf-Ta-Ti-Nb-V)C and (Ta-Ti-Nb-V-W)C high-entropy carbides, and then compositing them with the 19.2 vol% Co binder to make hard metals (also known as cemented carbides) as potential cutting tool materials [9]. Researchers also mix up the TiC and FeCoNiCuAl HEA with mechanical alloying and sinter the mixture into TiC-reinforced FeCoNiCuAl HEA composites with spark plasma sintering, whose hardness and wear performance are largely improved [10]. Lastly, a low-cost bulk metallic glass, $Zr_{58}Cu_{15.46}Ni_{12.74}Al_{10.34}Nb_{2.76}Y_{0.5}$ (at%), is prepared using the industrial grade Zr and its high-cycle fatigue behavior is investigated [11].

3. Conclusions and Outlook

Though eight articles were eventually accepted for publication, this Special Issue covers a wide range of topics in the field of high-entropy materials, the majority of which cover different aspects of high-entropy alloys, while high-entropy ceramics, high-entropy composites, and bulk metallic glasses are also covered. The published work aims to inspire continued research in the respective sub-fields. We look forward to seeing more interesting and inspiring research following this Special Issue, advancing each sub-field and the wider field of high-entropy materials as a whole.

This Special Issue could not be a success without the contribution of committed authors, responsive reviewers, and the diligent *Metals* Editorial Office, especially Toliver Guo, who facilitates everything with his patience and professionalism. Thanks to all authors, reviewers, and *Metals* staff for their dedication. Credit goes to every single person engaged in this endeavor.

Funding: This research received no external funding.

Conflicts of Interest: The authors declare no conflict of interest.

References

1. Yeh, J.W.; Chen, S.K.; Lin, S.J.; Gan, J.Y.; Chin, T.S.; Shun, T.T.; Tsau, C.H.; Chang, S.Y. Nanostructured high-entropy alloys with multiple principal elements: Novel alloy design concepts and outcomes. *Adv. Eng. Mater.* **2004**, *6*, 299–303. [CrossRef]
2. Li, W.; Liaw, P.K.; Gao, Y. Fracture resistance of high entropy alloys: A review. *Intermetallics* **2018**, *99*, 69–83. [CrossRef]
3. Li, W.; Xie, D.; Li, D.; Zhang, Y.; Gao, Y.; Liaw, P.K. Mechanical behavior of high-entropy alloys. *Prog. Mater. Sci.* **2011**, *118*, 100777. [CrossRef]
4. Regenberg, M.; Hasemann, G.; Wilke, M.; Halle, T.; Krüger, M. Microstructure Evolution and Mechanical Properties of Refractory Mo-Nb-V-W-Ti High-Entropy Alloys. *Metals* **2020**, *10*, 1530. [CrossRef]
5. Panina, E.; Yurchenko, N.; Zherebtsov, S.; Stepanov, N.; Salishchev, G.; Ventzke, V.; Dinse, R.; Kashaev, N. Laser Beam Welding of a Low Density Refractory High Entropy Alloy. *Metals* **2019**, *9*, 1351. [CrossRef]
6. Zhang, Y.; Zhang, M.; Li, D.; Zuo, T.; Zhou, K.; Gao, M.C.; Sun, B.; Shen, T. Compositional Design of Soft Magnetic High Entropy Alloys by Minimizing Magnetostriction Coefficient in $(Fe_{0.3}Co_{0.5}Ni_{0.2})_{100-x}(Al_{1/3}Si_{2/3})_x$ System. *Metals* **2019**, *9*, 382. [CrossRef]
7. Tsai, M.-H.; Tsai, R.-C.; Chang, T.; Huang, W.-F. Intermetallic Phases in High-Entropy Alloys: Statistical Analysis of their Prevalence and Structural Inheritance. *Metals* **2019**, *9*, 247. [CrossRef]
8. Castro, D.; Jaeger, P.; Baptista, A.C.; Oliveira, J.P. An Overview of High-Entropy Alloys as Biomaterials. *Metals* **2021**, *11*, 648. [CrossRef]
9. Pötschke, J.; Dahal, M.; Vornberger, A.; Herrmann, M.; Michaelis, A. Production and Properties of High Entropy Carbide Based Hardmetals. *Metals* **2021**, *11*, 271. [CrossRef]
10. Zhu, T.; Wu, H.; Zhou, R.; Zhang, N.; Yin, Y.; Liang, L.; Liu, Y.; Li, J.; Shan, Q.; Li, Q.; et al. Microstructures and Tribological Properties of TiC Reinforced FeCoNiCuAl High-Entropy Alloy at Normal and Elevated Temperature. *Metals* **2020**, *10*, 387. [CrossRef]
11. Zhou, S.; Zhang, T.; Li, L.; Yang, J.; Zhang, M.; Wang, C.; Zhang, Y. Fatigue Behavior of $Zr_{58}Cu_{15.46}Ni_{12.74}Al_{10.34}Nb_{2.76}Y_{0.5}$ Bulk Metallic Glass Fabricated by Industrial-Grade Zirconium Raw Material. *Metals* **2021**, *11*, 187. [CrossRef]

metals

Review

An Overview of High-Entropy Alloys as Biomaterials

Diogo Castro [1], Pedro Jaeger [2], Ana Catarina Baptista [1] and João Pedro Oliveira [3,*]

1 CENIMAT/I3N, Departamento de Ciência dos Materiais, Faculdade de Ciências e Tecnologia, FCT, Universidade Nova de Lisboa, 2829-516 Campus de Caparica, Portugal; ddp.castro@campus.fct.unl.pt (D.C.); anacbaptista@fct.unl.pt (A.C.B.)
2 Instituto de Soldagem e Mecatrônica LABSOLDA, Universidade Federal de Santa Catarina—UFSC, Florianópolis 88040-900, Brazil; pedro.jaeger@posgrad.ufsc.br
3 UNIDEMI, Department of Mechanical and Industrial Engineering, NOVA School of Science and Technology, Universidade NOVA de Lisboa, 2829-516 Caparica, Portugal
* Correspondence: jp.oliveira@fct.unl.pt

Abstract: High-entropy alloys (HEAs) have been around since 2004. The breakthroughs in this field led to several potential applications of these alloys as refractory, structural, functional, and biomedical materials. In this work, a short overview on the concept of high-entropy alloys is provided, as well as the theoretical design approach. The special focus of this review concerns one novel class of these alloys: biomedical high-entropy alloys. Here, a literature review on the potential high-entropy alloys for biomedical applications is presented. The characteristics that are required for these alloys to be used in biomedical-oriented applications, namely their mechanical and biocompatibility properties, are discussed and compared to commercially available Ti6Al4V. Different processing routes are also discussed.

Keywords: high-entropy alloys; biomedical materials

check for updates

Citation: Castro, D.; Jaeger, P.; Baptista, A.C.; Oliveira, J.P. An Overview of High-Entropy Alloys as Biomaterials. *Metals* **2021**, *11*, 648. https://doi.org/10.3390/met11040648

Academic Editors: Weidong Li and Leszek Adam Dobrzanski

Received: 3 March 2021
Accepted: 31 March 2021
Published: 15 April 2021

1. Introduction

High-entropy alloys are loosely defined as alloys containing at least five principal elements, each having an atomic percentage ranging between 5 and 35%, according to Yeh et al. [1], who alongside Cantor et al. [2], first discovered these alloys. There are other proposed parameters and definitions, such as the non-existence of intermetallic phases on a solid solution that would make an high-entropy alloy [3]. For these materials, high entropy refers to the high value of mixing entropy between equimolar metal alloys, conventionally, more than 1.5 R, resulting in a high probability of forming random solid solutions during solidification instead of forming intermetallic compounds or segregated states. Similarly, a medium-entropy alloy [4] is defined as a material composed of two to four principal elements or having a mixing entropy between R and 1.5 R [5].

The origin of studies related to high-entropy alloys can be traced back to the 1980s though the first published works with actual alloys only came out in 2004, as previously described [1,2]. These pioneer works marked the beginning of an increasing number of studies regarding this new branch of alloy development [6]. Among some of the key properties exhibited by these alloys in their early years, one can account for their excellent mechanical behavior, the high corrosion resistance, and the suggestion of the existence of effects such as the high-entropy effect, lattice distortion effect, and slow diffusion effect [1].

Ranganathan [7] described the limitations on the development of multicomponent alloys, since the amount of information needed to study these alloys as the number of components increase was difficult to manage, and because by adding processing variables, it would be of astronomical proportions. Although this turned out to be true, in recent years, there has been an increase in the number of ways currently available for the fast and reliable identification of new alloys compositions of interest.

High-entropy alloys are also known to possess unique microstructural features giving rise to improved properties over conventional engineering alloys. There are several contributions to the extraordinary properties of high-entropy alloys. These contributions are known as the four core effects, as described by Yeh [8].

Figure 1 represents the relationships between composition, structure and microstructure, and physical and mechanical properties of high-entropy alloys, providing the context in which the four core effects take place [6,8–11].

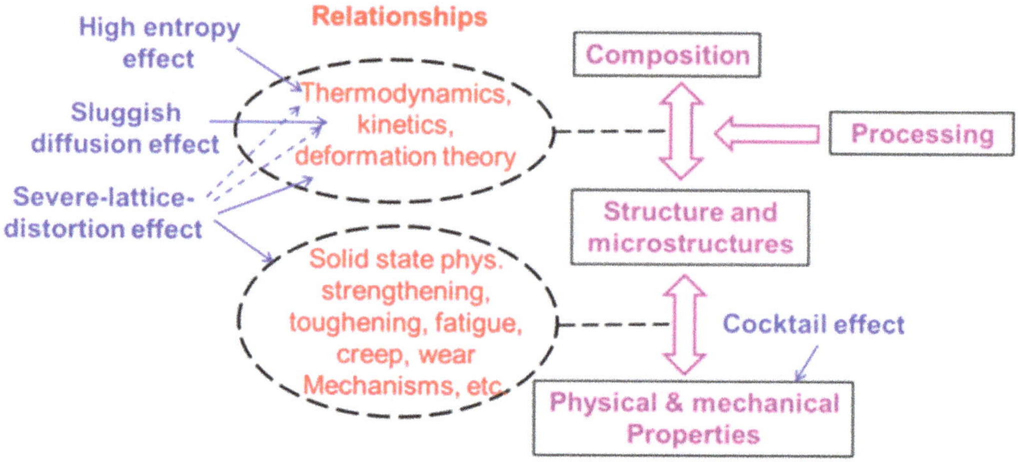

Figure 1. The scheme of physical metallurgy showing areas that are influenced by the four core effects exhibited by high-entropy alloys (from [6]). Reprinted with permission from Springer Nature: eBook "High-Entropy Alloys-Fundamentals and Applications", Copyright 2021, License: 5021540456140.

The four core effects are as follows: high-entropy effect; sluggish diffusion; severe lattice distortion; and cocktail effect. The high-entropy effect is where, contrary to what was previously thought to be the case, mixing equimolar chemically compatible elements narrows down the possible solid solution phases of an alloy. The high mixing entropy in multiprincipal elements lowers the Gibbs free energy for random and partially ordered solid solutions to compete with intermetallic compounds on attaining equilibrium states at high temperatures. On the other hand, having high-entropy alloys with a more significant difference in the enthalpy of mixing between unlike atom pairs, leads to the formation of more than two phases [12]. The atomic size difference parameter (δ) also plays a role in this and is compared alongside the entropy of mixing (ΔS_{mix}) and enthalpy of mixing (ΔH_{mix}) in describing trends, setting phase selection rules [13], and describing the order–disorder competition in these alloys. The severe lattice distortion effect, where different atomic sizes, bonding energies, and crystal structure tendencies among the constituent elements all contribute to the lattice distortion, typically results in a large solution hardening effect. This effect leads to enhanced properties such as strength and hardness, and the temperature dependance of these properties is also consequently low. The sluggish diffusion effect, where the number of principal elements in the system's matrix is proportional to the degree of sluggish diffusion, leads to higher recrystallization temperatures and creep resistances, slower grain growth and particle coarsening, and ease of getting supersaturated states and fine precipitates. Finally, the cocktail effect is where each phase, being a multiprincipal element solid solution, is regarded as both an atomic-scale and microscale composite, having multiprincipal element composite effects and multiphase composite effects, respectively. Due to the interactions between the different elements and the severe lattice distortion, the rule of mixtures of the different elements underpredicts the alloys' changes by a substantial amount.

1.1. Alloy Design Strategy: The Theoretical Approach

Parameters such as enthalpy of mixing, entropy of mixing, melting points, atomic size difference, and valence electron concentration are used as phase formation rules for high-entropy alloys. These rules and parameters allow researchers to predesign a high-entropy alloy with the characteristics that best suit their envisaged application.

1.1.1. Enthalpy and Entropy of Mixing

To understand these two concepts, a couple of assumptions are required to be established. When equal atom size and loose atomic packing are assumed, the configuration entropy of mixing of an alloy is equivalent to that of an ideal gas. Moreover, there are competing states in the solid state of an alloy, which are elemental phases, intermetallic compounds, and solid solution phases (random solid solutions and partially ordered solid solutions). The one with the lowest free energy at a given temperature and pressure will be predominant in the equilibrium state, and so, if kinetic factors are not involved, phase formation is thermodynamically controlled by the Gibbs free energy, G:

$$\Delta G_{mix} = \Delta H_{mix} - T \times \Delta S_{mix} \quad (1)$$

Therefore, the configurational entropy of mixing, under said assumptions, is defined as follows:

$$\Delta S_{mix} = -R \sum_{i=1}^{n} x_i \times \ln(x_i) \quad (2)$$

with R being the gas constant and x_i the molar fraction of the ith element, and the enthalpy of mixing for multicomponent alloys [13] is defined as follows:

$$\Delta H_{mix} = \sum_{i=1, i \neq j}^{n} 4\Delta H_{AB}^{mix} x_i x_j \quad (3)$$

where ΔH_{AB}^{mix} is the enthalpy of mixing for a binary equiatomic AB alloy and x_i and x_j are the molar fractions of the ith element.

The enthalpy of mixing is proportional to the number of bonds of unlike atoms, and completely ordered structures have a higher number of unlike atom pair bonds than that of random solid solutions, and although this difference lowers with the increase in the number of principal elements in an equiatomic high-entropy alloy, the number of unlike atom pair bonds in random solid solutions will never be the same or higher than that of intermetallic compounds. The reason a solid solution manages to compete with intermetallic compounds is because the entropy of mixing for completely ordered structures is ≈ 0, while in disordered structures such as random solid solutions, the ΔS_{mix} equals $R \ln n$ [10].

Of course, equal atom size and loose atomic packing cannot be assumed for real alloys, and so the above definition has its limits. For this, another parameter is used, considering atomic composition, the atom size, and the overall packing density. According to Mansoori et al. [14], the total configurational entropy of mixing S_T per sphere for a hard sphere system is defined as follows:

$$S_T(c_i, r_i, \xi) = S_C(c_i) + S_E(c_i, r_i, \xi) \quad (4)$$

where $S_C = -k_B \sum_{i=1}^{n} c_i \ln c_i$ is the equivalent to ΔS_{mix}, the configurational entropy of mixing for an ideal solution. S_E is a function of the atomic composition, c_i, the atom size r_i, and the overall packing density, ξ, and is defined as the excessive entropy of mixing [14–16]. Thus, for high values of ξ, non-equiatomic compositions, and different atomic sizes between constituent elements, there is a correction provided by S_E that makes S_T a more precise value of configurational entropy of mixing, whereas for high temperatures, ξ is negligible enough so that ΔS_{mix} remains a good approximation. The variation of the total entropy of mixing, S_T, with the atomic packing fraction, ξ, for different high-entropy alloys detailed in Figure 2. It is worth noting that with S_T, the difference in the total entrop of mixing for different high-entropy alloy systems with the same number of princip elements, becomes increasingly evident when atomic packing fraction gets higher wit decreasing temperature.

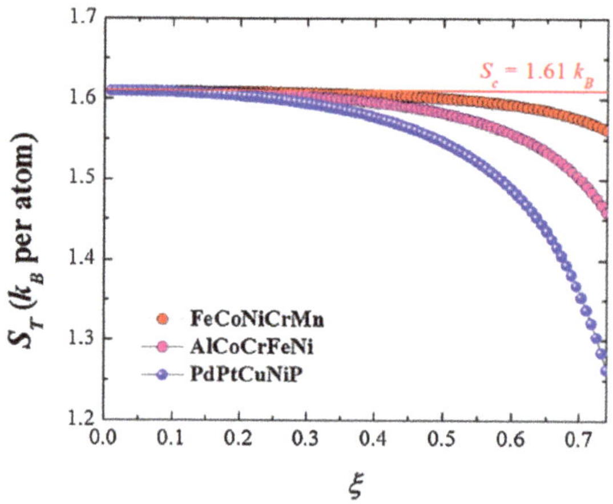

Figure 2. Variation of the total entropy of mixing, S_T, with the atomic packing fraction j for three fiv component alloys of equiatomic composition (from [9]). Reprinted with permission from Elsevie Mater. Today. 19, Copyright 2021, License: 5047200903826.

1.1.2. Ω Parameter

Zhang et al. [17] proposed the Ω parameter, which combines the effects of ΔS_{mix} an ΔH_{mix} on the stability of a multicomponent solid solution, taking the various component' melting points into consideration. The Ω and T_m parameters are defined by

$$\Omega = \frac{T_m \times \Delta S_{mix}}{|\Delta H_{mix}|} \quad (5$$

$$T_m = \sum_{i=1}^{N} x_i (T_m)_i \quad (6$$

where T_m is the average melting temperature of the N-element alloy and $(T_m)_i$ is th melting point of the ith component of the alloy.

Similar to how Yeh et al. [1] suggested $\Delta S_{mix} > 1.6R$ as a criterion for high-entrop alloys, Zhang et al. proposed $\Omega > 1.1$, where the higher the Ω value, the higher th probability of forming a single-phase random solid solution in high-entropy alloys. Th parameter's potential for such estimation was originally intended to be used along wit the atomic size difference parameter, δ, which is described below.

1.1.3. Atomic Size Difference

The atomic size effect has been thoroughly corroborated and defined by severa experimental data throughout the literature [17], and the critical atomic size difference (δ found for a solid solution to form high-entropy alloys is $\delta \leq 6.6\%$. δ is defined as [13]

$$\delta = \sqrt{\sum_{i=1}^{N} x_i \left(1 - \frac{d_i}{\sum_{j=1}^{N} x_j d_j}\right)^2} \quad (7$$

where N is the number of elements, x_i or x_j is the composition of the ith or jth element, and d_i or d_j is the atomic diameter of the ith or jth component. In Figure 3, we can see the discussed parameters used and paired with each other, having several phase overlaps. Some regions are identified, but ultimately, no definitive separation between them is achieved.

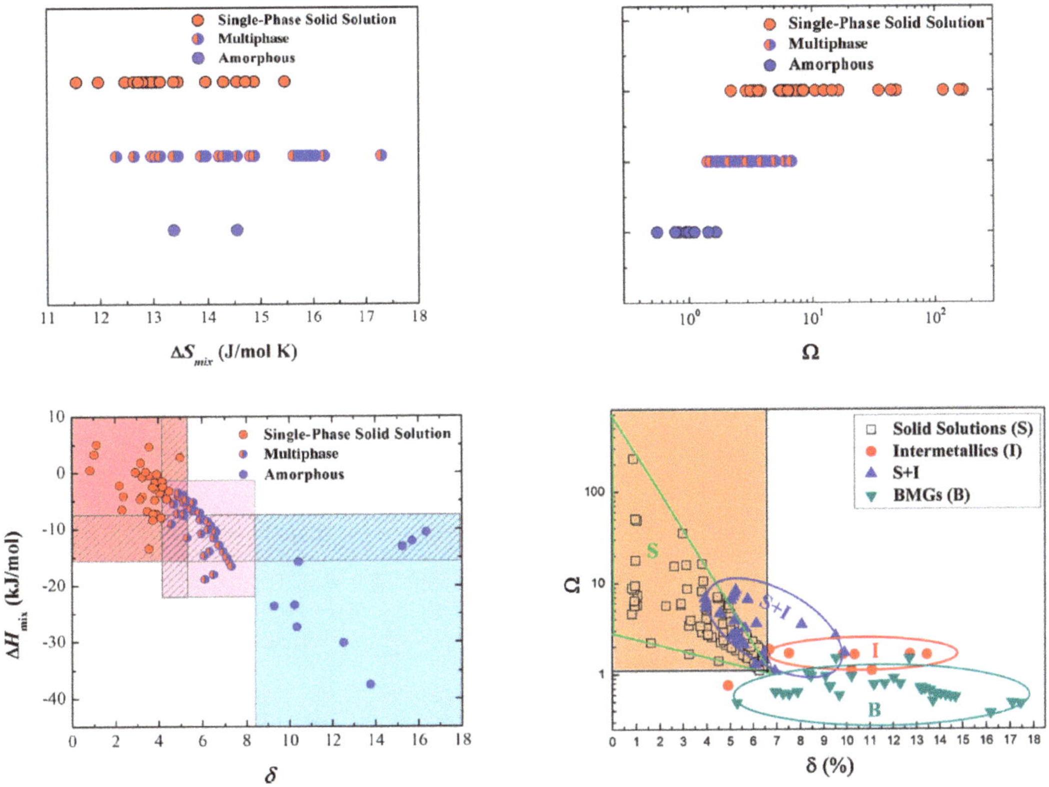

Figure 3. ΔS_{mix}, Ω, and δ vs. ΔH_{mix} (from [9]) and δ vs. Ω (from [17]) plots for as-cast high-entropy alloys (HEAs) with different phases, showing several phase overlaps. Reprinted with permission from Elsevier: Mater. Today. 19 (from [9]) and Mater. Chem. Phys. (from [17]), Copyright 2021, License: 5047200903826 and 5033770520802, respectively.

1.1.4. Φ Parameter

Ye et al. [16] pieced together the individual parameters S_C and S_E along with T_m and ΔH_{mix}, proposing the ϕ parameter, defined as

$$\phi = \frac{S_C - |\Delta H_{mix}|/T_m}{|S_E|} \tag{8}$$

Both ΔS_{mix} and Ω had significant overlaps in their respective values computed for the high-entropy alloys with different phases, and ϕ has shown to be the best out of the three parameters, with a critical value of $\phi_c = 20$, where alloys with higher values display single-phase solid solutions, and lower values display multiphase and amorphous solutions. Figure 4 shows the single parameter being used to successfully separate the different phase formations.

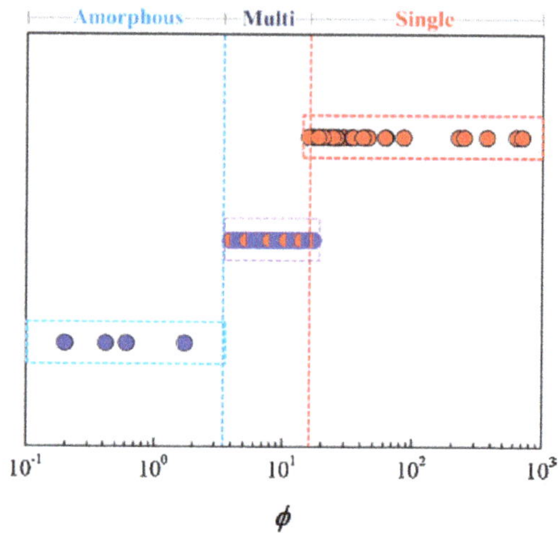

Figure 4. ϕ plot for different high-entropy alloy phases (from [9]). Reprinted with permission from Elsevier: Mater. Today. 19, Copyright 2021, License: 5047200903826.

1.1.5. Valence Electron Concentration

As described earlier, we have high entropy, causing these alloys to have mainly random solid solutions instead of intermetallic compounds, and we have a set of parameters that allows the prediction of whether they form single or multiple phases. The valence electron concentration (VEC) allows the determination of the type of structure that will be formed in high-entropy alloys, namely face-centered cubic (FCC), body-centered cubic BCC, or hexagonal close packed (HCP) structures. Figure 5 pairs both the ϕ parameter and the VEC, successfully separating both phase formation and crystal structure within reasonable and usable margins.

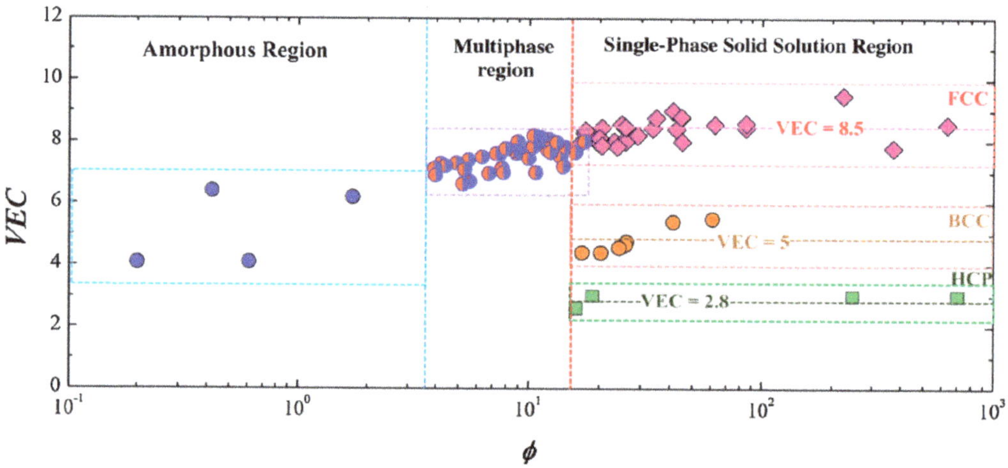

Figure 5. ϕ vs. valence electron concentration (VEC) plot for different high-entropy alloys (from [9]). Reprinted with permission from Elsevier: Mater. Today. 19, Copyright 2021, License: 5047200903826.

Different mean values of VEC have been found to regulate the crystal structure of existing phases in these alloys. Typically, for FCC structures, the VEC is ≈8.5, while for

BCC and HCP structures, it decreases to ≈5 and ≈2.8, respectively. When the d-electrons in the valence band are included, the average VEC of an alloy can be defined as:

$$\mathrm{VEC} = \sum_{i=1}^{n} c_i (VEC)_i \tag{9}$$

where $(VEC)_i$ is the VEC for the ith element [18].

Not only is VEC a good indicator of the predominant structure for an alloy, but there are also correlations with some mechanical properties such as the ductility of conventional BCC alloys. For instance, Mo- and W-based binary alloys can be ductile if VEC decreases [18,19]. Sheikh et al. [20] reported a similar rule for high-entropy alloys, indicating a possible VEC value threshold of 4.5, where alloys with lower values would lead to high ductility, and VEC > 5 would lead to high brittleness. These rules were studied for refractory high-entropy alloys, comprising group IV (Ti, Zr, Hf), V (V, Nb, Ta), and VI (Cr, Mo, W) refractory elements.

2. Biomedical Spectrum of High-Entropy Alloys

It may come as a surprise how, after years and years of developing science and technology, we are still to find an absolute replacement for bones and joints. The mechanical properties required to withstand the constant and cyclic loads of simple daily activities such as walking without fracture are not so simple, and to top it off, the biocompatibility requirements for a material to be used as an implant are such that the possibilities are further narrowed down. Therefore, materials used in such applications must have a low modulus of elasticity, to avoid stress shielding [21]; high yield strength; high fatigue resistance; and high ductility to be able to handle loads from physical activity. Natural load bearing joints such as the knee or hip joints are extremely well lubricated, with the bone tissue having friction coefficients of 0.003–0.0015, which are very low, and last for over 70 years [22], and so the tribological properties of implants are of utmost importance. In total hip arthroplasty (THA), wear is seen as a major cause of osteolysis, leading to bone loss due to the foreign-body autoimmune response, caused by the implant's wear debris [23,24]. Therefore, high surface-level wear and corrosion resistance (accordingly to the contact with the tissue or body fluid) and a low coefficient of friction are also essential, besides the obvious strict and demanding biocompatibility needs [25,26]. There are a series of standards for evaluating the biocompatibility of medical devices. These standards are part of ISO 10993 [27], serving as a framework for the development of biocompatibility evaluations. Such standards are recognized worldwide, with little additional (or fewer) recommendations from regulatory agencies such as the Food and Drug Administration (FDA) and the Japanese Ministry of Labor Health and Welfare (MHLW).

The International Organization for Standardization details the necessary tests for each application, and biomedical application devices based on high-entropy alloys are mainly focused on load bearing joints, e.g., in knee and hip arthroplasty, and so the proper categorization falls into implant devices, with contact to tissue/bone with a type C duration (more than 30 days) [28]. Of course, this list does not mandate any testing whatsoever, hence the name "Endpoints to be addressed in a biological risk assessment". It simply provides the biological effects to be taken into consideration given the class of the medical device being tested. Having said this, the associated testing can be quite slow and expensive (e.g., subacute/subchronic toxicity tests) and requires careful planning and a finished device ready for market. This means that early development stages such as material design and validation do not justify long and expensive tests. There are, however, more economically and logistically reasonable tests that can be carried out in early development. In vitro tests such as cell proliferation, live/dead viability assays, osteoblast adhesion and morphology, and bio-corrosion behaviors with potentiodynamic polarization curves in Hank's or phosphate buffer solutions are common, and some studies also include in vivo tests in mice, which cover more biological effects than the in vitro ones.

There are several alloys already used as biomedical implants. One of the most use alloys is Ti6Al4V [29], also known as Ti64, and thus, it is frequently used for compariso when alternatives are being sought. Although common, there is still room for improvemer in both its mechanical and tribological properties, stress shielding, and the presence c cytotoxic elements inherent to Ti6A4lV. Such disadvantages are being tackled by the deve opment of novel biomedical high-entropy alloys, where several mechanical and tribologic properties can be improved dramatically, while retaining excellent biocompatibility.

Since 2004, there has been an exponential increase in the number of published pape on high-entropy alloys [6,8–11,30,31]. Although most of the literature pertaining to hig entropy alloys is focused on structural application under a wide range of environment it did not take long until applications for medical implants began to be considered. Nex some recent developments in this field are described and discussed.

2.1. *Comparison of Mechanical Properties for Bone Implants*

First, we have biocompatibility factors limiting the elements that are usable in th material, e.g., the requirement of high yield strength and ductility to support heavy an cyclic loads during long-term usage, corrosion, and wear resistance and the need for a lo Young's modulus to avoid stress shielding, as detailed above.

Below, a list of high-entropy alloys studied for possible biomedical applications i presented (refer to Table 1). In this list, only high-entropy alloys that already had thei mechanical properties determined (i.e., Young's modulus, yield strength, and ductility) ar presented. Given the biocompatibility of the constituent elements in Table 1, an efficienc function (i) is presented for the mechanical properties and is obtained considering the rati between a material's resilience modulus U_r, and $(E - E_{Bone})$, where E is the alloy's Youn modulus and E_{Bone} is the corresponding elastic modulus of the bone. To simplify, E_{Bone} wi be regarded as the average value of 20 GPa [25]. This is an attempt to rank and order th data for an approximation of the desired and ideal mechanical properties. Thus, we defin

$$U_r = \frac{(Yield\ strength)^2}{2E} \quad (10)$$

$$i = \frac{\frac{(Yield\ strength)^2}{2E}}{(E - E_{Bone})} \quad (11)$$

It is worth noting that the simplicity of such a function and its linearity will not facto in any diminishing contributions either from yield stress or from Young modulus above o below ideal values, for the appropriate application, meaning it will not effectively measur the mechanical performance within the necessary and reasonable parameters, since thos parameters differ with each biomedical application scenario.

For instance, selective laser melted $Ti_{1.4}Nb_{0.6}Ta_{0.6}Zr_{1.4}Mo_{0.6}$ [32] displays a yiel strength of about 1690 MPa, which for some applications might prove to be unnecessaril high, in detriment to some other alternative with less, yet sufficient, yield strength an ductility, but with better (i.e., lower) Young modulus values to avoid stress shielding, fo example, for young patients whose bone will need to regenerate at an acceptable rate fo far longer. Table 1 summarizes the mechanical properties of some of the recently develope high-entropy alloys targeting biomedical applications.

Table 1. Known materials for medical implants and some recently developed biomedical high-entropy alloys.

Material	Condition	E (GPa)	Yield Strength (MPa)	Ductility (%)	Efficiency Function *i* (10^{-5})
Ti-6Al-4V (ELI) *	Mill annealed [25,33]	110	850–900	10–15	3.87
Equimolar TiZrNbTaMo	As-cast arc-melted [34]	153	1390 ± 75	$\varepsilon_p \approx 6$	4.75
$Ti_{1.4}Nb_{0.6}Ta_{0.6}Zr_{1.4}Mo_{0.6}$	Selective laser melting [32]	140 ± 9	1690 ± 78	ε_t = 1.32 ± 0.19	8.50
Equimolar TiZrHfNbTa	As-cast arc-melting in a Ti-gettered high-purity Ar atmosphere [35]. Remelted at least 6 times to ensure chemical homogeneity	103	1200	12	8.42
TiZrNbTa (Alloy 1)		89	970	23	7.66
$Ti_{6.3}Zr_{6.3}Nb_{6.3}Ta$ (Alloy 2)		75	790	29	7.56
$Ti_7Zr_7Nb_5Ta$ (Alloy 3)		69	780	26	9.00
Ti_9Zr_9NbTa (Alloy 4)		57	690	24	11.29
$TiZrNbTa_{1.62}$ (Alloy 5)		93	1050	12.7	8.12
$TiZrNb_{2.33}Ta_{2.33}$ (Alloy 6)		135	970	12.5	3.03
$Ta_{0.7}HfZrTi$		77	1046	14.31	12.46
$Ta_{0.8}HfZrTi$		85	1120	4.63	11.35
$(TaNb)_{0.6}HfZrTi$		73	880	19	10.01

* ELI = Extra Low Interstitials, referring to grade 23 of Ti-6Al-4V.

Not all high-entropy alloys from the work done by Yuan et al. [35] are referred in Table 1. Only the ones with more promising mechanical characteristics were chosen, although it is worth noting that the high-entropy alloys with HCP or HCP + BCC structures in Yuan's work all exhibited a relatively poorer mechanical performance if we take all the three characteristics (Young modulus, yield strength, and ductility) into consideration. The values for yield strength and ductility are obtained from tensile tests in the case of Yuan et al. work. Ti6Al4V is also presented for comparison, and we can see that apart from Alloy 6 [35], all other high-entropy alloys present a better efficiency function *i*, and most of them also have a better overall performance if you factor in ductility. There is no constituent element in these HEAs that has any cytotoxicity individually.

In the TiZrNbTa system, high values of Ti and Zr not only led to low values of the Young's modulus but also had lower yield strength compared to alloys 1 and 5. High values of Nb and Ta (alloy 6) had a sharp increase in Young modulus with only a slight decrease in yield strength.

In the $Ta_xHfZrTi$ system, a single BCC structure was observed only for x = 0.7, 0.8, and 1, and the last two had lower ductility (<5%). As such, $Ta_{0.7}HfZrTi$ was the alloy with the highest *i* value, having a good combination of Young modulus and yield strength.

From the $(TaNb)_xHfZrTi$ system, where x = 0, 0.2, 0.4, 0.5, 0.6, 0.8, and 1, which was experimentally tested in Yuan et al. [35], only x = 0.6 and x = 1 are in the table, since all other conditions had higher Young's modulus, similar yield strengths, and lower ductility (except for x = 0.2, which had 32.5%).

Both selective laser melted $Ti_{1.4}Nb_{0.6}Ta_{0.6}Zr_{1.4}Mo_{0.6}$ and equimolar TiZrNbTaMo have shown a rather small ductility (ε_t = 1.32% and $\varepsilon_p \approx$ 6%, respectively) compared to the rest of the materials in Table 1 (excluding the $Ta_{0.8}HfZrTi$ configuration). Wang et al. [34] suggests that brittleness and limited plasticity under deformation warrants further development of this system, away from equiatomic compositions.

More developments that have been occurring in this field, but that are not included in Table 1, are discussed below.

Gurel et al. [36] explored the TiTaHfNb, TiTaHfNbZr, and TiTaHfMoZr systems in terms of mechanical properties and fracture behavior under impact loading (Charpy V-notch test). All three BCC-structured alloys show a relatively lower Young's modulus when compared to more conventional implants. TiTaHfNb high-entropy alloy (composition of $Ti_{77.11}Ta_{6.54}Hf_{6.63}Nb_{12.73}$, in at.%) had the highest energy absorption capacity and highest ductility, out of the three systems, and under impact loading, significant difference has been observed in the dislocation activity in the microstructure when adding both Zr and Mo to the system. Both elements had negative impact in terms of the energy absorption and brittleness of the TiTaHfNb and TiTaHf alloy systems. With a composition of $Ti_{77.11}Ta_{6.54}Hf_{6.63}Nb_{12.73}$, this high-entropy alloy has a measured Young modulus of 112.2 GPa, a Charpy V-notched average impact energy of 14.8 J, and a hardness value of 3.5 GPa (using a Berkovich type indenter).

The TiTaHfNbZr system was also studied by Yang et al. [37]. An equiatomic composition processed by usual vacuum arc melting under a Ti-gettered Ar atmosphere, was tested for its biocorrosion behavior in a simulated physiological environment and in vitro biocompatibility. It showed a passivating behavior with low passive current density of about 0.08×10^{-2} A/m^2 (0.07×10^{-2} A/m^2 for Ti6Al4V), a corrosion rate of 5.6×10^{-4} mm/year (5.9×10^{-4} mm/year for Ti6Al4V), and high electrochemical impedance in Hank's solution at 310 K. MTT assay results for MC3T3-E1 cell-proliferation activities on the TiZrHfNbTa HEA and Ti-6Al-4V alloy after 1, 3, 5, and 7 days of incubation show that the high-entropy alloy has a slightly better cell viability, although not statistically significant. The authors did not mention this, but as seen in this review, there are several alternatives to this system and composition that reveal better biocompatibility when compared to the commercial Ti6Al4V, not to mention the high cost of constituent elements such as Hf.

A variant of the TiTaHfNbZr high-entropy alloy, $Ti_{1.5}ZrTa_{0.5}Hf_{0.5}Nb_{0.5}$ (with the equimolar sample named RHEA1 (refractory high entropy alloy) and the non-equimolar one being RHEA2) has been experimentally tested, and its properties were compared to that of the conventional 316 L stainless steel, CoCrMo, and Ti6Al4V alloys, namely in terms of wear resistance, wettability, pitting, and general corrosion resistance [38]. With a continuous passive plateau up to 1800 $mV_{Ag/AgCl}$ without the pitting effect and more positive E_{corr} and lower I_{corr} than that of Ti6Al4V (which starts gradually exhibiting pitting corrosion behavior at 863.0 $mV_{Ag/AgCl}$), this alloy shows considerably higher pitting corrosion resistance and general corrosion resistance. Compared to its equimolar counterpart, the $Ti_{1.5}ZrTa_{0.5}Hf_{0.5}Nb_{0.5}$ has a significantly thicker corrosion-resistant barrier oxide film, which is attributed to a higher content of electronegative elements such as Ti and Zr and also a lower lattice strain (0.0037, compared to the 0.0306 from the equimolar counterpart, which is connected to the higher amount of differently sized elements such as Hf, Nb, and Ta). The contact angles measured in the sessile drop test with phosphate buffered saline (PBS) for 316 L, CoCrMo, Ti6Al4V, RHEA1, and RHEA2 were 75.4, 73.3, 56.9, 59.5, and 61.9°, respectively, showing a similar wettability with Ti6Al4V, and a superior one to 316 L and CoCrMo. Its elastic modulus (E) and hardness (H) were 98.57 and 3.02 GPa, respectively. This hardness value was taken after a pile-up correction made using AFM 3D imaging after nanoindentation tests. This pile-up was observed to be higher in intensity on the conventional materials than on both high-entropy alloys. The wear resistance was inferred trough H/E and H^3/E^2 ratios, shown in Table 2 and found to be better than the 316 L, CoCrMo, and Ti6Al4 materials. With a superior wear resistance, similar wettability, lower Young's modulus, and significantly better corrosion resistance, this high-entropy alloy system's configuration ($Ti_{1.5}ZrTa_{0.5}Hf_{0.5}Nb_{0.5}$) has shown promising potential in the biomedical field and warrants further investigation into its biocompatibility.

Table 2. Mechanical properties of 316 L, CoCrMo, Ti6Al4V, RHEA1 (TiZrTaHfNb equimolar sample) and RHEA2 (Ti1.5ZrTa0.5Hf0.5Nb0.5 non-equimolar sample) measured by nanoindentation after applying a pile-up correction (from [38]). Reprinted with permission from Elsevier: Intermetallics, Copyright 2021, License: 5044280969444.

Alloy	H (GPa)	E (GPa)	H/E	H^3/E^2
316 L	2.44 ± 0.18	195.62 ± 9.32	0.012	0.00038
CoCrMo	4.21 ± 0.15	222.72 ± 6.70	0.019	0.0015
Ti6Al4V	3.32 ± 0.15	128.20 ± 8.14	0.026	0.0022
RHEA 1	3.14 ± 0.12	112.74 ± 4.66	0.028	0.0024
RHEA 2	3.02 ± 0.11	98.57 ± 4.18	0.031	0.0028

Based on the TiNbTaZr system, Todai et al. [39] designed a TiNbTaZrMo alloy from several TiNbTaZrX alternatives (X = Cr, V, Mo, W, Fe), using some parameters in the literature (ΔS_{mix}, ΔH_{mix}, δ, Ω, and VEC). The authors found that the Cr, V, and Fe options had a large atomic size difference, from the δ parameter. Mo and W, however, had a high possibility for the formation of a solid solution, and Mo was eventually chosen for the fact it was selected in conventional metallic biomaterials [40–44]. The TiNbTaZrMo system was similar in its elemental segregation in the DR/IR (dendrite/inter-dendrite) zones both in its as–cast and annealed form, but both mechanical and biocompatibility tests have shown what was stated to be promising potentials for HEAs in the biomedical sphere. Next are two studies on this system, discussing possible property trends in composition shifts.

Nagase et al. [45] explored the TiNbTaZr and TiZrHfMo systems, designing and producing $TiZrHfCr_{0\cdot2}Mo$ and $TiZrHfCo_{0\cdot07}Cr_{0\cdot07}Mo$ systems (arc melted), and comparing them to TiNbTaZr and TiNbTaZrX (X = V, Mo and W) in terms of hardness, and to 316 L, CoCrMo, and CP-Ti in terms of biocompatibility (cell density on osteoblast cultivation for 24 h). Both $TiZrHfCr_{0\cdot2}Mo$ and $TiZrHfCo_{0\cdot07}Cr_{0\cdot07}Mo$ were found to have a hardness of 531 and 532 HV, respectively, while having a lower liquidus temperature, T_L, than the TiNbTaZr and TiNbTaZrX counterparts. The cell density evaluations have shown that both the produced high-entropy alloys presented a lower, yet statistically similar, level of biocompatibility.

Akmal et al. [46] studied the $(MoTa)_xNbTiZr$ (TiZrNbTaMo) system for various values of x. A critical concentration of x = 0.4 was determined, where higher concentrations of Mo and Ta led to increasing brittleness and elastic modulus and decreasing ductility. The optimal concentration was found to be x = 0.2. The in vivo biocompatibility tests and the soft tissue response of the alloy (x = 0.2) showed no signs of toxic response in the performed tests. The electrochemical behavior in PBS at room temperature showed superior corrosion resistances relative to the commercial 316 L stainless steel, with passivity values of 1.2 V without signs of pitting.

Variants of the TiZrNbTaMo system were also studied by Hua et al. [47] with a molar ratio of 0.5, 1, 1.5, and 2 for the Ti concentration. These high-entropy alloys showed a dendritic structure with two BCC solid solution phases. The 0.5 Ti molar ratio alloy presented a hardness of about 500 HV and a compressive strength of nearly 2600 MPa, a plastic strain of over 30%, and the best corrosive wear resistance among the four alloys studied. This molar ratio had a rather coarsened dendrite structure relative to the rest in this study. The increase in Ti was also linked to the formation of short and round dendrite arms, attributed to the increasing fraction of the low-melting-temperature element. Corrosion wise, the alloys tested presented similar corrosion current densities in PBS solution and better corrosion potentials, meaning higher passivation film stability. XPS results show mainly Ti^{4+}, Zr^{4+}, Nb^{5+}, Ta^{5+}, Mo^{4+}, and Mo^{6+} oxides in the surface film of the HEAs. The literature indicates that high concentrations of Ti, Zr, and Nb can lead to passivation, protecting the bare metal from corrosion [38,48], and uniform distribution of Mo across different phases is also linked to dense and stable passivation films [49].

2.2. Coatings

Aside from producing bulk high-entropy alloys, some researchers have also devoted their time and efforts to develop these alloys via coating-based technologies. These are discussed now.

Braic et al. [26] took on TiZrNbHfTa-based system coatings, (TiZrNbHfTa)N and (TiZrNbHfTa)C, and investigated their biocompatibility. These were, as stated, the first high-entropy alloy coatings to be considered as biomaterials. The produced films were found to be composed of simple FCC solid solutions with a (111) preferred orientation. Simulated body fluid (SBF) testing revealed the carbide coating with the highest non-metal to metal ratio (carbon/metal ≈ 1.3) to have the highest hardness (≈31 GPa), best friction behavior (μ = 0.12), and the highest wear resistance. Biocompatibility tests have also shown good adherence from osteoblast cells to the coating with no cytotoxic responses and a very high ratio of live cells compared with those that were found dead for all the investigated groups.

In 2019, Tüten et al. [50] followed a previous study that first explored radio frequenc (RF) magnetron sputtering using TiTaHfNbZr equimolar targets on NiTi substrates, but di not the analyze wear and corrosion resistances [51]. Tüten studied the equimolar TiTaHfN bZr high-entropy alloy thin film deposited on a Ti6Al4V substrate by RF magnetro sputtering. The coating exhibited high hardness and elastic modulus (12.51 ± 0.34 GP and 181.3 ± 2.4 GPa, respectively), with enhanced results in wear resistance and coeffi cient of friction. The small grain size and uniform amorphous structure were attribute as the main strengthening mechanisms, corroborating previous studies on the hardnes enhancement in amorphous-structured films [52,53]. This study corroborated an invers trend observed in thin films, where increased surface roughness led to a lower coefficier of friction and wear rate, suggesting that further work must be done to address this.

Guo et al. [49] used a cluster-plus-glue-atom model to successfully design a refractor high-entropy alloy, $(TiZrNb)_{14}SnMo$ both in bulk (via vacuum arc melting) and as a coatin in a pure-Ti substrate (via laser cladding). The TiZrNb system used an average virtua element M as solvent, since they have weak interactions between them, and then used S and Mo as the cluster-center solute atom and glue atom, respectively, since out of the tw Sn has the strongest interactions with the average virtual element's constituents (Ti, Zr, an Nb). The bulk high-entropy alloy consisted of both BCC dendritic phase and rod-like HC (Sn, Zr) ordered phase. The coating had reduced elemental segregation between dendrit (DR) and inter-dendrite (IR), as expected, and so, this ordered phase was fully restraine in the as-clad coating. The coating had a lower Young's modulus (E_{DR} = 97.9 GPa an E_{IR} = 88.6 GPa) than its bulk counterpart (E_{DR} = 116.65 GPa and E_{IR} = 110.47 GPa Wear rate, corrosion, wettability, and cell adhesion tests were performed. Both bulk an coating outperformed Ti6Al4V on all those tests. For example, the coating's wear volum loss was less than a third of that of Ti6Al4V and with almost double the cell covering are ratio for the 7 days of cell culture for the growth morphology test of MG-63 cells, indicatin an excellent candidate for a biomaterial. Both the coating and bulk refractory high entrop alloys (RHEAs) revealed high hardness.

2.3. Elemental Homogenization

Alternative processing routes are also being explored to enhance the properties o high-entropy alloys via elemental homogenization of the microstructure, as will be dis cussed now.

Perumal et al. [54] used stationary friction processing (SFP) on a MoNbTaTiZr HEA managing to homogenize surface-level elemental distribution to a degree of efficiency not possible by annealing. In about 15 min, they achieved a level of elemental homoge nization across dendritic and inter-dendritic regions equivalent to an annealing treatmen at 1275 K for a whole week. Biocorrosion resistance tests have shown a near sixfold in crease compared to the as-cast counterpart, with a similar result for biocompatibility tests which indicates higher viability compared to both the as-cast counterpart and the mor conventional biomaterials such as 316 L stainless steel and Ti6Al4V.

In hopes to improve the microstructure's homogeneity, powder metallurgy (PM routes have also been tried [55]. Popescu et al. explored the TiZrNbTaFe system with PM method and further corroborated the findings that additions of Nb and Zr to a T matrix lead to the formation of a multicomponent alloy with an extremely high resistanc to active and passive dissolution in acids, as stated by Yu et al. [56]. The corrosion rate were notably better than that of Ti6Al4V, but direct comparisons to its as-cast counterpar was not experimentally made. It does, however, suggest some potential in TiZrNbTaF systems for biomedical applications. Although some useful information was found in thi work, there was no elemental analysis of the resulting microstructure to be able to compar the effect on segregation mitigation against other processing routes.

Selective laser melting has also received some attention for its high cooling rat characteristics [32,57]. Ishimoto et al. compared an as-cast and a selective laser melting buil $Ti_{1.4}Nb_{0.6}Ta_{0.6}Zr_{1.4}Mo_{0.6}$ high-entropy alloy [32]. The two differed both in microstructur and mechanical properties. The additively manufactured alloy had a very low level of elemental segregation compared to the as-cast counterpart. The dendrite structure was compared for chemical composition and results are shown in Table 3. Compared to the as-cast counterpart, the selective laser melted part also showed improved osteoblast adhesion, better compressive yield strength (1690 from 1140 MPa), and a true strain at break of 1.32 ± 0.19%, this was attributed to the very refined microstructure induced by the high cooling rate induced by selective laser melting.

Table 3. Comparison of the local chemical composition (at.%) of a selective laser melting (SLM) and as-cast $Ti_{1.4}Nb_{0.6}Ta_{0.6}Zr_{1.4}Mo_{0.6}$ biomedical high-entropy alloy (from [32]).

In at.%.		Ti	Nb	Ta	Zr	Mo
Nominal concentration		30.4	13.0	13.0	30.4	13.0
SLM	Dendrite	30.7	12.0	13.0	30.7	13.7
	Inter-dendrite	31.4	10.9	10.7	34.3	12.6
As-cast	Dendrite	26.8	16.6	22.6	17.3	16.9
	Inter-dendrite	30.8	11.5	8.66	38.3	10.6

2.4. *Biocompatibility Studies*

Thus far, most the evaluation of biomedical high-entropy alloys primarily focused on its mechanical properties and phase stability. However, some recent works have studied the in vivo and in vitro properties of these novel engineering alloys.

Akmal et al. studied the in vivo biocompatibility of alloys from the Mo–Ta–Nb–Ti–Zr system [46]. When their optimized alloy, the selection of which was made based on mechanical properties, was used as implant in mice, it was observed that no muscle damage or adverse effects due to potential toxicity problems occurred. Importantly, one of the key conclusions from Akmal et al.'s work concerns the fact that, for this alloy system, such impacts can be considered as safe as those of key biomedical materials such as NiTi shape memory alloys or Ti–6Al–4V alloys. Yang et al. evaluated the in vitro compatibility of a TiZrHfNbTa high-entropy alloy [37]. Good cell adhesion, viability, and proliferation of the MC3T3-E1 cells used in this study highlighted the potential of such an alloy to be used in biomedically oriented applications. Wang et al. studied the biocompatibility of a Ti40Zr20Hf10Nb20Ta10 high-entropy alloy for human gingival fibroblasts [58]. Overall, no cytotoxicity was demonstrated when the alloy was in contact with live tissue. Moreover, compared to its TiZrHf counterpart, the TiZrHfNbTa alloy exhibited improved proliferation of fibroblasts, which translates to better biocompatibility properties.

The lack of systematic studies on the biocompatibility of potential biomedical alloys is something that is expected to be addressed soon given the extraordinary properties of these materials. Up to now, these materials are primary candidates for biomedical applications but need further verification to establish their boundaries of application.

3. Conclusions

Several biomedical-oriented high-entropy alloys are now being developed, uncovering the vast pool of compositional possibilities. Existing ones still have a long way to go in their development, be it in processing method, composition, and even compatibility between components in the case of coating technologies. There are an increasing number of relevant parameters that help in the design of these materials. The entropy ΔS_{mix} and enthalpy ΔH_{mix} of mixing [1]; the atomic size difference, δ [13], S_T, S_E, and S_C [14,15]; the Φ parameter [16]; and VEC are still the most used, but cluster-plus-glue-atom models are also being successfully applied [49].

In this short overview, a range of mechanical properties of several biomedical high-entropy alloys were compared and ranked with a proposed efficiency function, *i*. Ti_9Zr_9NbTa ($i = 11.29$), $Ta_{0.8}HfZrTi$ ($i = 11.35$), and $Ta_{0.7}HfZrTi$ ($i = 12.46$) were at the top, with $Ta_{0.8}HfZrTi$ having a relatively lower ductility. By comparison, Ti6Al4V had an $i = 3.87$. It must be noted that this list is very short, since the amount of standardized information regarding basic mechanical properties such as yield strength, Young's modulus, and ductility are not yet abundant in the literature, since most studies focus on other pressing issues such as corrosion behavior and biocompatibility. Therefore, finding reliably comparable data is not elementary. It is to be noted, however, that most biomedical high-entropy alloys discussed here have superior mechanical behavior than the commercially available Ti6Al4V, 316 L stainless steel, and CoCrMo alternatives.

Rapid cooling seems to be a standard goal for most processing methods due to the advantages in suppressing elemental segregation [10,32,49]. Rapid cooling processes such as selective laser melting are receiving some attention for that reason, resulting in grain refinement and enhanced solid solution strengthening [57]. While coating production does not have the same cooling speed as those found during selective laser melting, interesting properties are still obtained in the materials obtained by these methods.

Other approaches have been made to mitigate elemental segregation, including the powder metallurgy route by Popescu et al. [55] and stationary friction processing by Perumal et al. [54]. It is worth nothing that the latter had biocorrosion resistance tests show a near sixfold increase compared to the as-cast counterpart (MoNbTaTiZr system). Therefore, elemental homogenization is indeed a relevant part in further improving both high-entropy alloys and their associated biomedical properties.

In summary, the future of biomedical high-entropy alloys is promising, though there is a need for a more in-depth analysis of the microstructure/properties relationships for these alloys.

Author Contributions: Conceptualization: J.P.O.; Investigation: D.C., P.J., A.C.B., J.P.O.; Writing—original draft preparation: D.C.; Writing—review and editing: A.C.B., J.P.O. All authors have read and agreed to the published version of the manuscript.

Funding: This research was funded by FCT-MCTES, grant numbers UIDB/50025/2020-2023 and UIDB/00667/2020.

Acknowledgments: This work was partially funded by national funds through the FCT—Fundação para a Ciência e a Tecnologia, I.P., under the scope of the project UIDB/50025/2020-2023. J.P.O. acknowledges Fundação para a Ciência e a Tecnologia (FCT—MCTES) for its financial support via the project UIDB/00667/2020 (UNIDEMI).

Conflicts of Interest: The authors declare no conflict of interest.

References

1. Yeh, J.-W.; Chen, S.-K.; Lin, S.-J.; Gan, J.-Y.; Chin, T.-S.; Shun, T.-T.; Tsau, C.-H.; Chang, S.-Y. Nanostructured High-Entropy Alloys with Multiple Principal Elements: Novel Alloy Design Concepts and Outcomes. *Adv. Eng. Mater.* **2004**, *6*, 299–303. [CrossRef]
2. Cantor, B.; Chang, I.; Knight, P.; Vincent, A. Microstructural development in equiatomic multicomponent alloys. *Mater. Sci. Eng. A* **2004**, *375-377*, 213–218. [CrossRef]
3. Otto, F.; Yang, Y.; Bei, H.; George, E. Relative effects of enthalpy and entropy on the phase stability of equiatomic high-entropy alloys. *Acta Mater.* **2013**, *61*, 2628–2638. [CrossRef]
4. Gali, A.; George, E. Tensile properties of high- and medium-entropy alloys. *Intermet.* **2013**, *39*, 74–78. [CrossRef]
5. Miracle, D.B.; Miller, J.D.; Senkov, O.N.; Woodward, C.; Uchic, M.D.; Tiley, J. Exploration and Development of High Entropy Alloys for Structural Applications. *Entropy* **2014**, *16*, 494–525. [CrossRef]
6. *High-Entropy Alloys*; Springer Science and Business Media LLC: New York, NY, USA, 2016.
7. Ranganathan, S. Alloyed pleasures: Multimetallic cocktails. *Curr. Sci.* **2003**, *85*, 1404–1406.
8. Yeh, J.-W. Recent progress in high-entropy alloys. *Eur. J. Control.* **2006**, *31*, 633–648. [CrossRef]
9. Ye, Y.; Wang, Q.; Lu, J.; Liu, C.; Yang, Y. High-entropy alloy: Challenges and prospects. *Mater. Today* **2016**, *19*, 349–362. [CrossRef]
10. Yeh, J.-W. Alloy Design Strategies and Future Trends in High-Entropy Alloys. *JOM* **2013**, *65*, 1759–1771. [CrossRef]
11. Tsai, M.-H.; Yeh, J.-W. High-Entropy Alloys: A Critical Review. *Mater. Res. Lett.* **2014**, *2*, 107–123. [CrossRef]
12. Tong, C.-J.; Chen, Y.-L.; Yeh, J.-W.; Lin, S.-J.; Chen, S.-K.; Shun, T.-T.; Tsau, C.-H.; Chang, S.-Y. Microstructure characterization of Al x CoCrCuFeNi high-entropy alloy system with multiprincipal elements. *Met. Mater. Trans. A* **2005**, *36*, 881–893. [CrossRef]
13. Zhang, Y.; Zhou, Y.J.; Lin, J.P.; Chen, G.L.; Liaw, P.K. Solid-Solution Phase Formation Rules for Multi-component Alloys. *Adv. Eng. Mater.* **2008**, *10*, 534–538. [CrossRef]
14. Mansoori, G.A.; Carnahan, N.F.; E Starling, K.; Leland, T.W. Equilibrium Thermodynamic Properties of the Mixture of Hard Spheres. *J. Chem. Phys.* **1971**, *54*, 1523–1525. [CrossRef]

15. Ye, Y.; Wang, Q.; Lu, J.; Liu, C.; Yang, Y. The generalized thermodynamic rule for phase selection in multicomponent alloys. *Intermet.* **2015**, *59*, 75–80. [CrossRef]
16. Ye, Y.; Wang, Q.; Lu, J.; Liu, C.; Yang, Y. Design of high entropy alloys: A single-parameter thermodynamic rule. *Scr. Mater.* **2015**, *104*, 53–55. [CrossRef]
17. Yang, X.; Zhang, Y. Prediction of high-entropy stabilized solid-solution in multi-component alloys. *Mater. Chem. Phys.* **2012**, *132*, 233–238. [CrossRef]
18. Guo, S.; Ng, C.; Lu, J.; Liu, C.T. Effect of valence electron concentration on stability of fcc or bcc phase in high entropy alloys. *J. Appl. Phys.* **2011**, *109*, 103505. [CrossRef]
19. Qi, L.; Chrzan, D.C. Tuning Ideal Tensile Strengths and Intrinsic Ductility of bcc Refractory Alloys. *Phys. Rev. Lett.* **2014**, *112*, 115503. [CrossRef] [PubMed]
20. Sheikh, S.; Shafeie, S.; Hu, Q.; Ahlström, J.; Persson, C.; Veselý, J.; Zýka, J.; Klement, U.; Guo, S. Alloy design for intrinsically ductile refractory high-entropy alloys. *J. Appl. Phys.* **2016**, *120*, 164902. [CrossRef]
21. Song, Y.; Xu, D.; Yang, R.; Li, D.; Wu, W.; Guo, Z. Theoretical study of the effects of alloying elements on the strength and modulus of β-type bio-titanium alloys. *Mater. Sci. Eng. A* **1999**, *260*, 269–274. [CrossRef]
22. Sreekanth, P.S.R.; Kanagaraj, S. Wear of Biomedical Implants. In *Tribology for Scientists and Engineers*; Metzler, J.B., Ed.; Springer: New York, NY, USA, 2013; pp. 657–674.
23. Zhu, Y.H.; Chiu, K.Y.; Tang, W.M. Review Article: Polyethylene wear and osteolysis in total hip arthroplasty. *J. Orthop. Surg.* **2001**, *9*, 91–99. [CrossRef]
24. Amstutz, H.C.; Campbell, P.; Kossovsky, N.; Clarke, I.C. Mechanism and clinical significance of wear debris-induced osteolysis. *Clin. Orthop. Relat. Res.* **1992**, *276*, 7–18. [CrossRef]
25. Long, M.; Rack, H. Titanium alloys in total joint replacement—a materials science perspective. *Biomaterials* **1998**, *19*, 1621–1639. [CrossRef]
26. Braic, V.; Balaceanu, M.; Braic, M.; Vladescu, A.; Panseri, S.; Russo, A. Characterization of multi-principal-element (TiZrNbHfTa)N and (TiZrNbHfTa)C coatings for biomedical applications. *J. Mech. Behav. Biomed. Mater.* **2012**, *10*, 197–205. [CrossRef]
27. CSA. New 10993–1 Evaluation and testing within a risk management process. *Int. Organ.* **2007**, 1–11.
28. International Organization for Standardization. ISO 10993–1:2018—Biological Evaluation of Medical Devices Part 1: Evaluation and Testing within a Risk Management Process. 2018. Available online: https://www.iso.org/standard/68936.html (accessed on 30 March 2021).
29. Oliveira, J.; Panton, B.; Zeng, Z.; Andrei, C.; Zhou, Y.; Miranda, R.; Fernandes, F.B. Laser joining of NiTi to Ti6Al4V using a Niobium interlayer. *Acta Mater.* **2016**, *105*, 9–15. [CrossRef]
30. Zhang, Y.; Zuo, T.T.; Tang, Z.; Gao, M.C.; Dahmen, K.A.; Liaw, P.K.; Lu, Z.P. Microstructures and properties of high-entropy alloys. *Prog. Mater. Sci.* **2014**, *61*, 1–93. [CrossRef]
31. Derimow, N.A.; Abbaschian, R. Liquid Phase Separation in High-Entropy Alloys—A Review. *Entropy* **2018**, *20*, 890. [CrossRef]
32. Ishimoto, T.; Ozasa, R.; Nakano, K.; Weinmann, M.; Schnitter, C.; Stenzel, M.; Matsugaki, A.; Nagase, T.; Matsuzaka, T.; Todai, M.; et al. Development of TiNbTaZrMo bio-high entropy alloy (BioHEA) super-solid solution by selective laser melting, and its improved mechanical property and biocompatibility. *Scr. Mater.* **2021**, *194*, 113658. [CrossRef]
33. Wang, K. The use of titanium for medical applications in the USA. *Mater. Sci. Eng. A* **1996**, *213*, 134–137. [CrossRef]
34. Wang, S.-P.; Xu, J. TiZrNbTaMo high-entropy alloy designed for orthopedic implants: As-cast microstructure and mechanical properties. *Mater. Sci. Eng. C* **2017**, *73*, 80–89. [CrossRef] [PubMed]
35. Yuan, Y.; Wu, Y.; Yang, Z.; Liang, X.; Lei, Z.; Huang, H.; Wang, H.; Liu, X.; An, K.; Wu, W.; et al. Formation, structure and properties of biocompatible TiZrHfNbTa high-entropy alloys. *Mater. Res. Lett.* **2019**, *7*, 225–231. [CrossRef]
36. Gurel, S.; Yagci, M.; Canadinc, D.; Gerstein, G.; Bal, B.; Maier, H. Fracture behavior of novel biomedical Ti-based high entropy alloys under impact loading. *Mater. Sci. Eng. A* **2021**, *803*, 140456. [CrossRef]
37. Yang, W.; Liu, Y.; Pang, S.; Liaw, P.K.; Zhang, T. Bio-corrosion behavior and in vitro biocompatibility of equimolar TiZrHfNbTa high-entropy alloy. *Intermetallics* **2020**, *124*, 106845. [CrossRef]
38. Motallebzadeh, A.; Peighambardoust, N.S.; Sheikh, S.; Murakami, H.; Guo, S.; Canadinc, D. Microstructural, mechanical and electrochemical characterization of TiZrTaHfNb and Ti1.5ZrTa0.5Hf0.5Nb0.5 refractory high-entropy alloys for biomedical applications. *Intermetallics* **2019**, *113*, 106572. [CrossRef]
39. Todai, M.; Nagase, T.; Hori, T.; Matsugaki, A.; Sekita, A.; Nakano, T. Novel TiNbTaZrMo high-entropy alloys for metallic biomaterials. *Scr. Mater.* **2017**, *129*, 65–68. [CrossRef]
40. Okazaki, Y.; Ito, Y.; Kyo, K.; Tateishi, T. Corrosion resistance and corrosion fatigue strength of new titanium alloys for medical implants without V and Al. *Mater. Sci. Eng. A* **1996**, *213*, 138–147. [CrossRef]
41. Lee, S.-H.; Hagihara, K.; Nakano, T. Microstructural and Orientation Dependence of the Plastic Deformation Behavior in β-type Ti-15Mo-5Zr-3Al Alloy Single Crystals. *Met. Mater. Trans. A* **2011**, *43*, 1588–1597. [CrossRef]
42. Lee, S.-H.; Todai, M.; Tane, M.; Hagihara, K.; Nakajima, H.; Nakano, T. Biocompatible low Young's modulus achieved by strong crystallographic elastic anisotropy in Ti–15Mo–5Zr–3Al alloy single crystal. *J. Mech. Behav. Biomed. Mater.* **2012**, *14*, 48–54. [CrossRef]
43. Chiba, A.; Kumagai, K.; Nomura, N.; Miyakawa, S. Pin-on-disk wear behavior in a like-on-like configuration in a biological environment of high carbon cast and low carbon forged Co–29Cr–6Mo alloys. *Acta Mater.* **2007**, *55*, 1309–1318. [CrossRef]

44. Hiromoto, S.; Onodera, E.; Chiba, A.; Asami, K.; Hanawa, T. Microstructure and corrosion behaviour in biological environmen of the new forged low-Ni Co?Cr?Mo alloys. *Biomaterials* **2005**, *26*, 4912–4923. [CrossRef]
45. Nagase, T.; Iijima, Y.; Matsugaki, A.; Ameyama, K.; Nakano, T. Design and fabrication of Ti–Zr-Hf-Cr-Mo and Ti–Zr-Hf-Co-Cr-M high-entropy alloys as metallic biomaterials. *Mater. Sci. Eng. C* **2020**, *107*, 110322. [CrossRef] [PubMed]
46. Akmal, M.; Hussain, A.; Afzal, M.; Lee, Y.I.; Ryu, H.J. Systematic Study of (MoTa)xNbTiZr Medium- and High-Entropy Alloys fo Biomedical Implants- In Vivo Biocompatibility Examinations. *J. Mater. Sci. Technol.* **2020**, *78*, 183–191. [CrossRef]
47. Hua, N.; Wang, W.; Wang, Q.; Ye, Y.; Lin, S.; Zhang, L.; Guo, Q.; Brechtl, J.; Liaw, P.K. Mechanical, corrosion, and wear properti of biomedical Ti–Zr–Nb–Ta–Mo high entropy alloys. *J. Alloy. Compd.* **2021**, *861*, 157997. [CrossRef]
48. Zhou, Q.; Sheikh, S.; Ou, P.; Chen, D.; Hu, Q.; Guo, S. Corrosion behavior of Hf0.5Nb0.5Ta0.5Ti1.5Zr refractory high-entropy i aqueous chloride solutions. *Electrochem. Commun.* **2019**, *98*, 63–68. [CrossRef]
49. Guo, Y.; Li, X.; Liu, Q. A novel biomedical high-entropy alloy and its laser-clad coating designed by a cluster-plus-glue-ator model. *Mater. Des.* **2020**, *196*, 109085. [CrossRef]
50. Tüten, N.; Canadinc, D.; Motallebzadeh, A.; Bal, B. Microstructure and tribological properties of TiTaHfNbZr high entropy allo coatings deposited on Ti 6Al 4V substrates. *Intermetallics* **2019**, *105*, 99–106. [CrossRef]
51. Motallebzadeh, A.; Yagci, M.B.; Bedir, E.; Aksoy, C.B.; Canadinc, D. Mechanical Properties of TiTaHfNbZr High-Entropy Allo Coatings Deposited on NiTi Shape Memory Alloy Substrates. *Met. Mater. Trans. A* **2018**, *49*, 1992–1997. [CrossRef]
52. Shu, F.; Wu, L.; Zhao, H.; Sui, S.; Zhou, L.; Zhang, J.; He, W.; He, P.; Xu, B. Microstructure and high-temperature wear mechanisr of laser cladded CoCrBFeNiSi high-entropy alloy amorphous coating. *Mater. Lett.* **2018**, *211*, 235–238. [CrossRef]
53. Greer, A.L.; Rutherford, K.L.; Hutchings, I.M. Wear resistance of amorphous alloys and related materials. *Int. Mater. Rev.* **2002**, 4 87–112. [CrossRef]
54. Perumal, G.; Grewal, H.S.; Pole, M.; Reddy, L.V.K.; Mukherjee, S.; Singh, H.; Manivasagam, G.; Arora, H.S. Enhanced Biocorrosio Resistance and Cellular Response of a Dual-Phase High Entropy Alloy through Reduced Elemental Heterogeneity. *ACS Appl. B Mater.* **2020**, *3*, 1233–1244. [CrossRef]
55. Popescu, G.; Ghiban, B.; A Popescu, C.; Rosu, L.; Truscă, R.; Carcea, I.; Soare, V.; Dumitrescu, D.; Constantin, I.; Olaru, M.T.; et a New TiZrNbTaFe high entropy alloy used for medical applications. *IOP Conf. Ser. Mater. Sci. Eng.* **2018**, *400*, 022049. [CrossRe
56. Yu, S.Y.; Scully, J.R.; Vitus, C.M. Influence of Niobium and Zirconium Alloying Additions on the Anodic Dissolution Behavior c Activated Titanium in HCl Solutions. *J. Electrochem. Soc.* **2001**, *148*, B68–B78. [CrossRef]
57. Mohammed, M.T.; Semelov, V.G.; Sotov, A.V. SLM-built titanium materials: Great potential of developing microstructure an properties for biomedical applications: A review. *Mater. Res. Express* **2019**, *6*, 122006. [CrossRef]
58. Wang, S.; Wu, D.; She, H.; Wu, M.; Shu, D.; Dong, A.; Lai, H.; Sun, B. Design of high-ductile medium entropy alloys for denta implants. *Mater. Sci. Eng. C* **2020**, *113*, 110959. [CrossRef] [PubMed]

metals

MDPI

Article

Intermetallic Phases in High-Entropy Alloys: Statistical Analysis of their Prevalence and Structural Inheritance

Ming-Hung Tsai [1,2,*], Ruei-Chi Tsai [1], Ting Chang [1] and Wen-Fei Huang [1]

1 Department of Materials Science and Engineering, National Chung Hsing University, Taichung 402, Taiwan; trc19930408@gmail.com (R.C.T.); ace1123581321@gmail.com (T.C.); iamsophiahere@gmail.com (W.F.H.)
2 High Entropy Materials Center, National Tsing Hua University, Hsinchu 30013, Taiwan
* Correspondence: mhtsai@nchu.edu.tw; Tel.: +886-4-2285-1243

Received: 20 January 2019; Accepted: 16 February 2019; Published: 19 February 2019

Abstract: Strengthening high entropy alloys (HEAs) via second phases is a very effective approach. However, the design of intermetallic (IM) phases in HEAs is challenging, mainly because our understanding of IM phases in HEAs is still very limited. Here, a statistical approach is used to enhance our understanding towards IM phases in HEAs. A database consisting of 142 IM-containing HEAs was constructed. Our aim is twofold. The first is to reveal the most common IM phase types in published HEAs. The second is to understand whether HEAs inherit their IM structures from their binary/ternary subsystems, or whether they tend to form new structures irrelevant to their subsystems. The results show that the five most prevalent IM structures in the HEAs surveyed here are Laves, σ, B2, $L1_2$, and $L2_1$. This trend is evidently different from the overall trend among known binary/ternary IMs. As for structural inheritance, all the IM phases contained in the alloys are existing structures in the binary/ternary subsystems of the respective alloys. This suggests that the compositional complexity in HEAs does trigger additional complexity in IM structure formation. These findings have important implications in the future design and development of HEAs.

Keywords: high-entropy alloys; intermetallic; alloy design; phase stability

---

1. Introduction

High-entropy alloys (HEAs) have attracted significant attention in the past decade [1–3]. The formation of simple solid solutions (SSSs) in these complex alloys is unexpected and intriguing. Thus, a lot of effort in the HEA community has been devoted to the design and physical metallurgy of SSS HEAs [4–11]. Studies have shown that these alloys do have unusual properties [12–16]. However, their room temperature mechanical properties are often not as pleasing. In particular, the strengths of FCC alloys are often insufficient [17,18].

A very effective approach for strengthening alloys is the incorporation of second phases. Therefore, recently, some researchers proposed that instead of focusing on SSS alloys, more attention should be paid to HEAs that contain intermetallic (IM) phases [2,19,20]. In fact, incorporating second phases was even considered one of the three key strategies for the future development of HEAs [21]. Indeed, experimental results have shown that IM phases can improve the mechanical properties of HEAs. For example, the outstanding wear resistances in the $Al_{0.2}Co_{1.5}CrFeNi_{1.5}Ti$ and $Al_{0.3}CrFe_{1.5}MnNi_{0.5}$ alloys are the results of the formation of η and σ phases, respectively [22,23]. Very recently, it has been demonstrated that through carefully designed second phases, high strength and good ductility can even be achieved simultaneously [24–27]. For instance, the $(FeCoNiCr)_{94}Ti_2Al_4$ alloy shows a yield strength of over 1 GPa, while keeping a high tensile ductility of 17%. This is due to the engineering of γ' precipitates through various thermomechanical processes [24].

However, the tailoring of IM phases in HEAs for property enhancement is still very challenging. This is because compared with SSS phases; our fundamental understanding of IM phases in HEAs is still very limited. For example, there have been more than 20 publications regarding the formation criterion and stability of SSS phases in HEAs [6,28–31], although some of the earlier works have limited accuracy due to imprecise experimental data. In contrast, papers discussing IM phases in HEAs are very rare. Currently there are only one criterion on the general selection of IM phase type [32], and three stability criterions on two specific IM phase types (σ and Laves phases) [33–35].

To predict the stability of IM phases in HEAs, the reasonable first step is to identify the types of phases that could form in the alloy, and then compare the relative stabilities of these candidates. However, even the first step is difficult. Unlike the case in SSS phases, where only three types of structures are concerned (FCC, BCC, and HCP), there are more than two thousand structure types for IM phases [36]. A plausible assumption is that the candidate IM phases in HEAs are the IM phases in their subsystems (for quinary HEAs this means all quaternary, ternary, and binary subsystems). This is similar to the case in many ternary systems, where the IM phases are all inherited from the three binary subsystems. Nevertheless, even in ternary alloys this assumption is not necessarily true (i.e., there could be new structures that do not exist in the three binary subsystems). In HEAs, the composition is apparently more complex, which renders the above assumption even more questionable. In addition, in HEAs a new class of IM phases called high entropy intermetallics (HEI) have been reported [21]. These new IM phases show completely new compositions that have not reported before, which further complicates the determination of the candidate IM phases.

This work is the first part of an attempt to understand the IM phases in HEAs from a macroscopic viewpoint. A database of 142 IM-containing HEAs is constructed. Care is taken to reduce alloys with similar composition so that the diversity in the database is enhanced. The alloys and their respective IM structures are statistically analyzed. Our goal is twofold. The first is to reveal the most common IM phase types in HEAs. The results in HEAs are then compared with the general trends in binary and ternary IM phases and discussed. The second is to understand the inheritance of IM structures in HEAs. The IM structure types in the HEAs are compared with those in their subsystems to see if the structures are inherited from the subsystems of the alloys. The frequency which irrelevant (i.e., not existing in corresponding subsystems) structure types occur is also probed and discussed.

2. Construction of the Alloy Database

The HEAs in this research were collected using the following principles:

(1) Only alloys that contain IM phases were collected because our target is the IM phases.

(2) Alloys that have unidentified or uncertain IM phase types were excluded. In other words, either the formula (e.g., Fe_2Ti) or the common/trivial name (e.g., σ) of all the compounds in the alloy has to be reported. This is because including these alloys leads to difficulty in our analysis.

(3) Alloys with similar composition and identical IM phase constituents were listed only once (i.e., only one of the alloys was included). Two alloys are considered to have similar compositions if they have the same composing elements, and only the relative concentrations of the elements are different. This is because many studies fabricate series of alloys by systematically changing the content of a particular element. For example, He et al. [37] conducted detailed study on the effect of Al on the structure and properties of $Al_xCoCrFeMnNi$ alloys. They fabricated a total of 14 alloys (x = 0.02, 0.03, 0.04, 0.08, 0.09, 0.1, 0.11, 0.12, 0.14, 0.15, 0.16, 0.18, 0.19, and 0.25). Ten of the 14 alloys have the same IM phase constituent: B2. This similarity is not surprising; the alloys are composed of the same elements and the relative concentrations of these elements do not change significantly. If these alloys are all included, they and the IM phase they contain will have an unreasonably high weighting in the database, which will lead to biased results in subsequent analyses. Therefore, only one of the ten alloys was included in the database.

(4) If the same alloy contains different IM phases under different processing conditions, each condition with different IM phase constituents will be recorded separately.

However, in subsequent analyses these phases will be combined into one entry and counted only once. Assume the phase constituent of an alloy in the as-cast, 1100 °C annealed, and 900 °C annealed states are (α, β), (α, β, γ), and (α, δ), respectively. These three states will all be recorded in the database. However, when calculating, for example, the number of occurrences for each IM phase, the three records will be combined. The alloy will be regarded as containing four phases (α, β, γ, and δ) and counted only once.

Based on the above principles, a total of 142 IM-containing HEAs with sufficient compositional distinction are collected. These principles enhance the diversity of our database, and can markedly reduce the bias from compositional repetition. As an example, Miracle and Senkov published a comprehensive review on HEAs recently [3]. In their review, an HEA database with a total of 408 unique alloys was constructed. 247 of the 408 alloys contain IM phases. However, using the above principles, only 74 of the 247 alloys (~30%) will be included our database—70% of the alloys are excluded because of compositional similarity. This clearly shows the efficacy of the above principles.

3. Results and Discussion

3.1. Characterization of the Database

Table 1 shows the alloy database constructed in this research. It lists the composition, processing condition, and the solid solution and IM phases of each alloy. The abbreviations used in Table 1 are explained in the footnote of the table. Because the target of this research is the IM phase, each IM phase in the alloy is listed in a separate row. Three types of structure information are recorded (if possible). The first is the phase name. This is the common name for each structure (if it exists). Common name are either based on Strukturbericht notation or common trivial names [38] for certain structure types (most notably the topologically closed packed phases). For example, B2 is the Strukturbericht notation for cP2-CsCl and σ is the trivial name for tP30-CrFe. The second is the compound name reported in the original work (if reported). The third is the structure prototype of the phase. In many cases, the original publication only reports the common name of the phase without indicating the exact compound. In such cases the "compound reported" column will be left empty. Contrarily, if only the compound is reported, the common name of the phase will be identified based on the structure prototype. Moreover, to probe the structural inheritance in the alloys, the binary and ternary subsystems of each alloy are examined using various handbooks [39–41] to see whether the observed IM phases already exist in the subsystems. Binary subsystems are examined first. If the target IM phase can be found in any of the binary subsystems, the search is stopped and the subsystem will be recorded in the "IM Structure Found in" column. If the IM phase does not exist in any binary subsystem, the search will be continued in ternary ones and then recorded (if found). It should be noted that one IM phase can often be found in multiple subsystems. However, due to space limit only one of the subsystems is recorded.

Table 1. Composition, processing condition, phase constituents, and structural inheritance of the 142 IM-containing HEAs considered in this study.

Alloy Composition	Processing Condition [1]	Solution Phases [2]	IM Phases [3]			IM Structure Found in	Ref
			PhaseName	Compound Reported	Structure Prototype		
AgAlCoCrCuNi	AC	F, B	B2	-	CsCl	Al-Ni	[45]
$Al_{0.5}BCoCrCuFeNi$	AC	F	C32	CrB_2	AlB_2	B-Cr	[46]
			$L1_2$	-	Cu_3Au	Al-Ni	
$AlB_{0.5}CoCrCuFeMoNiSiTi$	AC + 1173K	B	B2	-	CsCl	Al-Ni	[47]
$Al_{0.3}C_{0.1}CoCrFeNi$	AC	F	-	$(Cr, Fe)_7C_3$	Cr_7C_3/Mn_7C_3	C-Cr	[48]
	AC + 1273K/72h	F	$D8_4$	$(Cr, Fe)_{23}C_6$	$Cr_{23}C_6$	C-Cr	[48]
	AC + 973K/72h	F	-	$(Cr, Fe)_7C_3$	Cr_7C_3/Mn_7C_3	C-Cr	[48]
			$D8_4$	$(Cr, Fe)_{23}C_6$	$Cr_{23}C_6$	C-Cr	
$Al_{0.5}C_{0.2}Co_{0.3}CrFeNi$	MA + SPS(1273K)	F, B	$D8_4$	$Cr_{23}C_6$	$Cr_{23}C_6$	C-Cr	[49]
			B2	-	CsCl	Al-Ni	
AlCoCrCuFe	SPS(1173K)	F	B2	-	CsCl	Al-Co	[50]
			σ	-	CrFe	Cr-Fe	
$Al_4(CoCrCuFeMnNiTiV)_{96}$	AC	B	$D0_{22}$	Al_3Ti	Al_3Ti	Al-Ti	[51]
$AlCoCrCuFeMo_{0.2}Ni$	Sinter(773K/0.5h) + 1573K/2h	F, B	σ	-	CrFe	Cr-Fe	[52]
$Al_8Co_{17}Cr_{14}Cu_8Fe_{17}Mo_{0.1}Ni_{34.8}TiW_{0.1}$	AC	F	$L1_2$	-	Cu_3Au	Al-Ni	[53]
$Al_{0.5}CoCrCuFeNi$	AC + CR + 1373K/5d	F-1, F-2, B	$L1_2$	-	Cu_3Au	Al-Ni	[54]
	AC + CR + 973K/5d	F-1, F-2, B	σ	-	CrFe	Cr-Fe	[54]
AlCoCrCuFeNi	AC	F	B2	-	CsCl	Al-Ni	[55]
			$L1_2$	-	Cu_3Au	Al-Ni	
$Al_{0.2}CoCrCu_{0.8}FeNiSi_{0.2}$	AC	F	$L1_2$	-	Cu_3Au	Al-Ni	[56]
$Al_{0.8}CoCrCu_{0.2}FeNiSi_{0.2}$	AC	B	B2	-	CsCl	Al-Ni	[56]
$Al_{0.5}CoCrCuFeNiTi$	AC	F, B-1, B-2	σ	CoCr	CrFe	Co-Cr	[57]
$Al_{0.5}CoCrCuFeNiTi_{1.2}$	AC	F, B-1, B-2	σ	CoCr	CrFe	Co-Cr	[57]
			N/A	Ti_2Ni	Ti_2Ni	Ni-Ti	
$Al_{0.5}CoCrCuFeNiTi_2$	AC	F, B-1, B-2	B2	-	CsCl	Al-Ni	[57]
			N/A	Ti_2Ni	Ti_2Ni	Ni-Ti	
$Al_{0.5}CoCrCuFeNiV$	AC	F, B	σ	-	CrFe	Cr-Fe	[58]
$Al_{0.4}Co_6Cr_{7.5}Cu_2Mn_{8.1}Ni_{32.2}Sn_{0.3}Ti_{12}V_{10}Zr_{21.5}$	AC		C14	-	$MgZn_2$	Al-Zr	[59]
			B2	TiNi	CsCl	Ti-Ni	
			N/A	Zr_7Ni_{10}	Zr_7Ni_{10}	Ni-Zr	
AlCoCrCuMnTi	AC	F, B-1, B-2	$L2_1$	-	$AlCu_2Mn$	Al-Cu-Mn	[60]
$AlCoCrCu_{0.5}Ni$	AC	F, B	B2	-	CsCl	Al-Ni	[61]
AlCoCrCuNiTiY	AC	F, B	N/A	Cu_2Y	KHg_2	Cu-Y	[62]
			$L2_1$	$AlNi_2Ti$	$AlCu_2Mn$	Al-Ni-Ti	
$Al_{0.15}CoCrFeMnNi$	AC	F, B	B2	-	CsCl	Al-Ni	[37]
$Al_{10}Co_{20}Cr_{10}Fe_{26}Mn_{14}Ni_{10}Ti_{10}$	AC + 973K/24h/WQ	F	χ	-	Ti_5Re_{24}	Ti-Cr-Fe	[34]
			$L2_1$	-	$AlCu_2Mn$	Al-Co-Ti	
$Al_{0.4}CoCrFeMnNiV$	AC + 973K/20h/AQ		σ	-	CrFe	Cr-Fe	[33]
$AlCoCrFeMo_{0.5}$	AC		B2	-	CsCl	Al-Co	[63]
			σ	-	CrFe	Cr-Fe	
$AlCoCrFeMo_{0.5}Ni$	AC		B2	-	CsCl	Al-Ni	[63]
			σ	-	CrFe	Cr-Fe	
$AlCoCrFeNb_{0.5}Ni$	AC	B	C14	$(Co, Cr)_2Nb$	$MgZn_2$	Co-Nb	[64]
$Al_{0.3}CoCrFeNi$	AC + CR(20%) + 1473K/0.5h/WQ + 823K/150h/WQ	F	$L1_2$	-	Cu_3Au	Al-Ni	[44]
	AC + CR(20%) + 1473K/0.5h/WQ + 973K/50h/WQ	F	B2	-	CsCl	Al-Ni	[44]
$Al_{0.3}CoCr_2FeNi$	AC + 973K/20h/AC	F, B	σ	-	CrFe	Cr-Fe	[33]
$Al_2CoCrFeNiSi$	CL	B	B2	-	CsCl	Al-Ni	[65]
$AlCo_{0.6}CrFe_{0.2}NiSiTi_{0.2}$	AC	B	A15	Cr_3Si	Cr_3Si	Cr-Si	[66]
$Al_4(CoCrFeNi)_{94}Ti_2$	AC + 1473K/4h + CR(30%) + 1273K /2h + 1073K/18h	F	$L1_2$	-	Cu_3Au	Al-Ni	[24]
			$L2_1$	-	$AlCu_2Mn$	Al-Ni-Ti	
$Al_{16.5}Co_{22.5}Cr_5Fe_{12.5}Ni_{30}Ti_{13.5}$	AC + 973K/24h/WQ	F	$L2_1$	-	$AlCu_2Mn$	Al-Ni-Ti	[34]
$Al_{0.5}CoCrFeNiTi_{0.5}$	AC	F, B	B2	-	CsCl	Al-Ni	[67]
			σ	-	CrFe	Cr-Fe	
$Al_{0.5}CoCrFeNiTi$	AC	B-1, B-2	B2	-	CsCl	Al-Ni	[68]
			C14	-	$MgZn_2$	Cr-Ti	

Table 1. *Cont.*

Alloy Composition	Processing Condition [1]	Solution Phases [2]	IM Phases [3]			IM Structure Found in	Ref
			PhaseName	Compound Reported	Structure Prototype		
$Al_{0.5}CoCrFe_2NiTi_{0.5}$	AC	F, B	B2	-	CsCl	Al-Ni	[67]
$AlCoCrFeNiTi_{1.5}$	AC	B	C14	Fe_2Ti	$MgZn_2$	Fe-Ti	[42]
	CL	F, B	N/A	Ti_2Ni	Ti_2Ni	Ni-Ti	[43]
$AlCoCrFeNiTi_2$	CL	F, B	B2		CsCl	Al-Ni	[43]
			N/A	Ti_2Ni	Ti_2Ni	Ni-Ti	
$AlCoCrFeNiZr_{0.5}$	AC	B	C15	-	$MgCu_2$	Co-Zr	[69]
			B2	-	CsCl	Al-Ni	
$Al_{0.5}CoCrNiTi_{0.5}$	AC + 1473K/24h	F, B	B2	-	CsCl	Al-Ni	[70]
	AC + 1373K/24h	F, B	B2	-	CsCl	Al-Ni	[70]
			σ	-	CrFe	Co-Cr	
	AC + 1173K/24h	F, B	σ	-	CrFe	Co-Cr	[70]
$Al_{2.5}CoCuFeNiSn_{0.1}$	AC	B	B2	-	CsCl	Al-Ni	[71]
			N/A	$Ni_{17}Sn_3$	Unkn.	Ni-Sn	
AlCoCuFeNiVZr	AC	F, B	N/A	Zr_2Ni_7	Zr_2Ni_7	Ni-Zr	[72]
AlCoCuFeNiZr	AC	F, B	N/A	$ZrFe_3Al$	Unkn.	Al-Fe-Zr	[73]
AlCoCuNiZn	SPS(873K/FC)	F	$L1_2$	-	Cu_3Au	Al-Ni	[74]
$AlCoFeMo_{0.5}Ni$	AC		B2	-	CsCl	Al-Ni	[75]
			σ	-	CrFe	Co-Mo	
$Al_{22}Co_{28}Ni_{31.75}Ti_{10.75}V_{7.5}$	AC + 973K/24h/WQ	F	B2	-	CsCl	Al-Ni	[34]
$Al_{15}Cr_{10}Co_{35}Ni_{35}Si_5$	AC	F	$L1_0$	-	CuAu	Al-Cr	[76]
			$L1_2$	-	Cu_3Au	Al-Ni	
$Al_{20}Cr_{25}Co_{25}Ni_{25}Si_5$	AC	B	B2	-	CsCl	Al-Ni	[76]
			σ	-	CrFe	Co-Cr	
AlCrCuFe	MA + SPS(1073K)	B	σ	-	CrFe	Cr-Fe	[77]
$AlCrCuFeMg_{0.5}$	MA + SPS(1073K)	B	σ	-	CrFe	Cr-Fe	[77]
			B2	AlFe	CsCl	Al-Fe	
			C15	$MgCu_2$	$MgCu_2$	Cu-Mg	
AlCrCuFeMg	MA + SPS(1073K)	B	σ		CrFe	Cr-Fe	[77]
			B2	AlFe	CsCl	Al-Fe	
			C15	$MgCu_2$	$MgCu_2$	Cu-Mg	
			C_b	Mg_2Cu	Mg_2Cu	Cu-Mg	
$AlCrCuFeMgNi_{4.75}$	MA + Sinter (1173K/2h)	-	$L1_2$	-	Cu_3Au	Al-Ni	[78]
$Al_{0.3}CrCuFeNi_2$	AC + CR(20%) + 1473K/0.5h/WQ + 823K/150h/WQ	F	$L1_2$	-	Cu_3Au	Al-Ni	[44]
$Al_{1.5}CrCuFeNi_2$	AC	F, B	B2	-	CsCl	Al-Ni	[79]
$AlCrCuFeNi_2$	AC	F, B	B2	-	CsCl	Al-Ni	[80]
			$L1_2$	-	Cu_3Au	Al-Ni	
AlCrCuFeNiTi	AC	B-1, B-2	C14	Fe_2Ti	$MgZn_2$	Fe-Ti	[81]
$Al_{14.1}Cr_6Fe_{28.2}Mn_{32.9}Ni_{18.8}$	AC	F	B2	-	CsCl	Al-Ni	[82]
$Al_{0.3}CrFe_{1.5}MnNi_{0.5}$	AC + 1373K/4h + 973K/2h/AQ	F	B2	-	CsCl	Al-Ni	[83]
			σ	-	CrFe	Cr-Fe	
$AlCrFe_2Ni_2$	AC	F, B	B2	-	CsCl	Al-Ni	[84]
AlCrFeMoNi	AC	B	σ	FeCrMo	CrFe	Cr-Fe	[85]
			B2	AlNi	CsCl	Al-Ni	
$Al_{15}Cr_5Fe_{30}Ni_{30}Ti_{20}$	AC + 973K/24h/WQ	-	χ	-	Ti_5Re_{24}	Ti-Cr-Fe	[34]
			C14	-	$MgZn_2$	Cr-Ti	
			$L2_1$	-	$AlCu_2Mn$	Al-Ti-Fe	
$Al_5Cr_{32}Fe_{35}Ni_{22}Ti_6$	AC	F, B	$L2_1$	-	$AlCu_2Mn$	Al-Ti-Fe	[86]
	AC + 1373K/6h/WQ + 1173K/1h/WQ	F	$L2_1$	-	$AlCu_2Mn$	Al-Ti-Fe	[86]
			σ	-	CrFe	Cr-Fe	
			η	–	Ni_3Ti	Ni-Ti	
$Al_{0.5}CrFeNiTiV$	AC + 973K/20h/AQ	-	$L2_1$	-	$AlCu_2Mn$	Al-Ti-Fe	[33]
			C15	-	$MgCu_2$	Cr-Ti	
$AlCr_{4.5}Mn_{13.6}Ni_{37.1}Sn_{0.3}Ti_{12}V_{10}Zr_{21.5}$	AC	-	C14	-	$MgZn_2$	Al-Zr	[87]
			C15	-	$MgCu_2$	Cr-Ti	
				TiNi	TiNi/CsCl	Ni-Ti	
AlCrMoNbTi	AC + 1373K/20h	B	C14	Cr_2Nb	$MgZn_2$	Cr-Nb	[88]
AlCrMoSiTi	AC	-	B2	-	CsCl	Al-Mo-Ti	[89]
			$D8_8$	Mo_5Si_3	Mn_5Si_3	Mo-Si	

Table 1. *Cont.*

Alloy Composition	Processing Condition [1]	Solution Phases [2]	IM Phases [3]			IM Structure Found in	Ref
			PhaseName	Compound Reported	Structure Prototype		
$Al_{0.5}CrNbTi_2V_{0.5}$	AC + 1473K/24h/AQ	B	C14	-	$MgZn_2$	Cr-Nb	[90]
AlCrNbTiVZr	AC + 1473K/24h	B	C14-1	ZrCrAl	$MgZn_2$	Cr-Nb	[91]
			C14-2	$ZrAl_2$	$MgZn_2$	Cr-Nb	
AlCrSiTiV coating on Ti-6Al-4V	CL	B	$D8_8$	Ti_5Si_3	Mn_5Si_3	Ti-Si	[92]
	CL + 1073K/24h	B	$D8_8$	Ti_5Si_3	Mn_5Si_3	Ti-Si	[92]
			$D8_2$	Al_8V_5	Zn_8Cu_5	Al-V	
$AlCu_{0.2}Li_{0.5}MgZn_{0.5}$	AC	-	$D8_e$	$Mg_{32}(Al,Zn)_{49}$	$Mg_{32}(Zn_2Al)_{49}$	Al-Mg-Zn	[93]
AlCuTiYZr	AC	B-1, B-2, H	N/A	$AlTi_2$	Unkn.	Al-Ti	[94]
$Al_{13}Fe_{36}Mn_{33}Ni_{18}$	AC	F	B2	-	CsCl	Al-Ni	[95]
$Al_{13}Fe_{36}Mn_{33}Ni_{18}Ti_6$	AC	F	B2	-	CsCl	Al-Ni	[95]
$AlMo_{0.5}NbTa_{0.5}TiZr$	AC + 1673K/24h/FC	B	B2	-	CsCl	Al-Mo-Ti	[56]
AlNbTaTiZr	HT-DSC (1823K/2cycles)	B	N/A	Al_4Zr_5	Ti_5Ga_4	Al-Zr	[96]
$Al_2NbTi_3V_2Zr$	MA + SPS (1173K/2h)	B	$L1_2$	-	Cu_3Au	Al-Zr	[97]
			N/A	Ti_2ZrAl	Mg_3Cd	Al-Ti	
AlNbTiVZr	AC + 1473K/24h	B	C14		$MgZn_2$	Al-Zr	[98]
			N/A	Zr_2Al	$Co_{1.75}Ge$	Al-Zr	
$Al_{1.5}NbTiVZr$	AC + 1473K/24h + 1273K/100h	B	C14		$MgZn_2$	Al-Zr	[98]
			N/A	Zr_2Al	$Co_{1.75}Ge$	Al-Zr	
			σ	$AlNb_2$	CrFe	Al-Nb	
$B_{22}(CoCrFeNi)_{78}$	AC + 1023K/10min	F	C16	Fe_2B	$CuAl_2$	B-Fe	[99]
$B_{26}(CoCrFeNi)_{74}$	AC + 873K/10min	F	C16	Fe_2B	$CuAl_2$	B-Fe	[99]
			$D8_1$	Cr_5B_3	Cr_5B_3	B-Cr	
				FeB	FeB-b/TlI	B-Fe	
$C_{0.1}CoCrFeMnNi$	AC + 1473K/14h/AQ	F	$D8_4$	$Cr_{23}C_6$	$Cr_{23}C_6$	C-Cr	[100]
$C_{0.25}CoCrFeMnNi$	AC + 873K/14h/AQ	F		Cr_7C_3	Cr_7C_3/Mn_7C_3	C-Cr	[100]
	AC + 1273K/14h/AQ	F	$D8_4$	$Cr_{23}C_6$	$Cr_{23}C_6$	C-Cr	[100]
				Cr_7C_3	Cr_7C_3/Mn_7C_3	C-Cr	
$C_{0.1}Hf_{0.5}Mo_{0.5}NbTiZr$	AC	B	B1	CHf	NaCl	C-Hf	[101]
CoCrCuFeMnNiTiV	AC	F, B	σ		CrFe	Cr-Fe	[51]
CoCrCuFeNi	MA + SPS(1173K)	F-1, F-2	σ	-	CrFe	Cr-Fe	[50]
CoCrCuFeNiTi	AC	F	C14	Fe_2Ti	$MgZn_2$	Fe-Ti	[102]
CoCrCuFeNiZr	AC		$C15_b$	$Zr(Ni, Cu)_5$	Be_5Au	Cu-Zr	[103]
CoCrFeHfNi	AC	B	C36		$MgNi_2$	Cr-Hf	[32]
CoCrFeMnNbNi	CL	F	C14	Fe_2Nb	$MgZn_2$	Fe-Nb	[104]
$Co_{0.5}CrFeMn_{1.5}Ni$	AC + 973K/20h/AQ	F	σ	-	CrFe	Cr-Fe	[33]
CoCrFeMnNiV	AC + 1273K/24h	F	σ	-	CrFe	Cr-Fe	[105]
$CoCrFeMo_{0.85}Ni$	AC	F	σ	-	CrFe	Cr-Fe	[106]
			μ	-	W_6Fe_7	Co-Mo	
	AC + 1273K/6d/WQ	F	μ	-	W_6Fe_7	Co-Mo	[107]
CoCrFeMoNi	AC	F	σ	-	CrFe	Cr-Fe	[32]
$Co_{1.5}CrFeMo_{0.5}Ni_{1.5}Ti_{0.5}$	AC	F	σ	-	CrFe	Cr-Fe	[108]
$CoCrFeNb_{0.3}Ni$	AC	F	C14	-	$MgZn_2$	Co-Nb	[109]
CoCrFeNbNi	AC	F	C14	-	$MgZn_2$	Cr-Nb	[32]
$CoCr_2FeNi$	AC + 973K/20h/AQ	F	σ	-	CrFe	Cr-Fe	[33]
CoCrFeNiTa	AC	F	C14	-	$MgZn_2$	Co-Ta	[32]
$CoCrFeNiTi_{0.3}$	AC	F, H	$L1_2$	-	Cu_3Au	Co-Ni	[110]
$CoCrFeNiTi_{0.5}$	AC	F	C36	Co_2Ti	$MgNi_2$	Co-Ti	[111]
			σ	CrFe	CrFe	Cr-Fe	
			R	$TiNi_2$	$TiNi_2$	Ni-Ti	
	AC + 1273K/6h/FC	F	σ		CrFe	Cr-Fe	[112]
			R	$TiNi_2$	$TiNi_2$	Ni-Ti	
CoCrFeNiTi	AC		χ		Ti_5Re_{24}	Ti-Cr-Fe	[32]
			C14		$MgZn_2$	Cr-Ti	
			η		Ni_3Ti	Ni-Ti	
	AC + 1273K/2h	F, B	N/A	$CoTi_2$	Ti_2Ni	Co-Ti	[113]
			B2	-	CsCl	Co-Fe	
$Co_{1.5}CrFeNi_{1.5}Ti$	AC + 1373K/4h + 1073K/10h/AQ	F	η	-	Ni_3Ti	Ni-Ti	[22]
CoCrFeNiV	AC	F	σ		CrFe	Cr-Fe	[32]
CoCrFeNiW	AC	F	μ		W_6Fe_7	Fe-W	[32]

Table 1. *Cont.*

Alloy Composition	Processing Condition [1]	Solution Phases [2]	IM Phases [3]			IM Structure Found in	Ref
			PhaseName	Compound Reported	Structure Prototype		
CoCrFeNiY	AC	B	N/A	YNi_3	$PuNi_3$	Ni-Y	[32]
			N/A	Y_2Ni_{17}	Th_2Ni_{17}	Ni-Y	
			B27	YNi	FeB-b	Ni-Y	
CoCrFeNiZr	AC	B	C15	-	$MgCu_2$	Co-Zr	[32]
CoCrMnNiV	AC + 1273K/3d	F	σ	-	CrFe	Co-Cr	[114]
CoCrNiTiV	MA + VHP (1423K/1h/FC)	F, B	σ	Ni_2V_3	CrFe	Ni-V	[115]
			η		Ni_3Ti	Ni-Ti	
$CoCuFeNiSn_{0.5}$	AC	F	N/A	$Cu_{81}Sn_{22}$	$Cu_{41}Sn_{11}$	Cu-Sn	[116]
CoCuFeNiTi	AC	F-1, F-2, B	C15	-	$MgCu_2$	Co-Ti	[117]
CoFeMnMoNi	AC + 1273K/3d	F	μ	-	W_6Fe_7	Co-Mo	[114]
$Co_{25}Fe_{25}Mn_{15}Ni_{15}Ti_{12.5}V_{7.5}$	AC + 973K/24h/WQ	F	C14	-	$MgZn_2$	Fe-Ti	[34]
CoFeMnNiV	AC + 1273K/3d	F	σ	-	CrFe	Fe-V	[114]
CoFeMnTiVZr	AC		C14	-	$MgZn_2$	Fe-Ti	[118]
CoFeMoNiV	AC	F	σ	Co_2Mo_3	CrFe	Co-Mo	[119]
$CoFeNb_{0.75}Ni_2V_{0.5}$	AC	F	C14	-	$MgZn_2$	Co-Nb	[120]
$CoFeNiSi_{0.5}$	AC	F	$D0_{11}$	Ni_3Si	Fe_3C	Ni-Si	[121]
$Co_{25}Fe_{25}Ni_{10}Ti_{30}V_{10}$	AC + 973K/24h/WQ		C14	-	$MgZn_2$	Fe-Ti	[34]
			C15	-	$MgCu_2$	Co-Ti	
$Co_{28.75}Fe_{31.25}Ni_{7.5}Ti_{30}V_{2.5}$	AC + 973K/24h/WQ		C14	-	$MgZn_2$	Fe-Ti	[34]
CoHfNbNiTiZr	AC	F, B	μ	NbNi	W_6Fe_7	Nb-Ni	[122]
	AC + 1073K/1h	F, B	μ	NbNi	W_6Fe_7	Nb-Ni	[122]
			N/A	$NbHf_2$	Unkn.	Hf-Nb	
$Co_2Mo_{0.8}Ni_2VW_{0.8}$	AC	F	μ	-	W_6Fe_7	Co-Mo	[123]
$CrFe_{1.5}MnNi_{0.5}$	AC	F	σ	CrFe	CrFe	Cr-Fe	[124]
CrFeMnNiTi	AC + 1273K/3d		χ	-	Ti_5Re_{24}	Ti-Cr-Fe	[114]
			C14	-	$MgCu_2$	Co-Ti	
CrFeNiTiVZr	AC + 1273K/24h	H	C14	-	$MgZn_2$	Fe-Ti	[125]
$CrFeNiV_{0.5}W$	AC	F, B	σ	-	CrFe	Cr-Fe	[126]
CrHfNbTiZr	AC + 1173K/10min	B	C15	-	$MgCu_2$	Cr-Hf	[127]
$Cr_{4.5}Mn_{13.6}Ni_{38.1}Sn_{0.3}Ti_{12}V_{10}Zr_{21.5}$	AC	-	C14	-	$MgZn_2$	Cr-Ti	[87]
			C15	-	$MgCu_2$	Cr-Ti	
			-	TiNi	TiNi/CsCl	Ti-Ni	
$CrNbMo_{0.5}Ta_{0.5}TiZr$	HIP(1723K/3h)	B-1, B-2	C15	-	$MgCu_2$	Cr-Ta	[128]
CrNbTaTiZr	HT-DSC (1823K/2cycles)	B, H	C14	-	$MgZn_2$	Cr-Nb	[96]
			C15	-	$MgCu_2$	Cr-Nb	
CrNbTiVZr	HIP(1473K/2h) + 1473K/24h	B	C15	-	$MgCu_2$	Cr-Nb	[129]
CrNbTiZr	HIP(1473K/2h) + 1473K/24h	B	C15	-	$MgCu_2$	Cr-Nb	[129]
CuFeMnNiPt	AC + 1373K/12h/FC		$L1_2$	-	Cu_3Au	Fe-Ni	[130]
$Hf_{0.5}Mo_{0.5}NbSi_{0.9}TiZr$	AC	B	$D8_8$	Zr_5Si_3	Mn_5Si_3	Si-Zr	[131]
$HfNbSi_{0.5}TiV$	AC	B	$D8_8$	Ti_5Si_3	Mn_5Si_3	Si-Ti	[132]
HfNbTiVZr	AC + 1173K/10min	B	C15	-	$MgCu_2$	V-Zr	[127]
$Mo_{0.7}NbTiV_{0.3}Zr$	AC + 1273K/72h	B-1, B-2	C15	-	$MgCu_2$	V-Zr	[133]
MoTaVWZr	AC	B-1, B-2, H	C15	-	$MgCu_2$	V-Ta	[134]
NbSnTaTiZr	AC	B	N/A	Sn_3Zr_5	Ti_5Ga_4	Sn-Zr	[96]
NbTiVZr	AC + 1273K/100h	B, H	C15	-	$MgCu_2$	V-Zr	[98]

[1] The abbreviations used for processing conditions are: AC = As-Cast; MA = Mechanical Alloyed; SPS = Spark Plasma Sintered; CR = Cold Rolled; WQ = Water Quenched; AQ = Air Quenched; CL = Cladded; FC = Furnace Cooled; HT-DSC = High Temperature-Differential Scanning Calorimeter; VHP = Vacuum Hot Press; HIP = Hot Isostatic Pressing. [2] The abbreviations used for SSS phases are: F = FCC; B = BCC; H = HCP. When there is more than one FCC or BCC phases, a number will be added after the abbreviation. For example, F-1 and F-2. [3] The abbreviations for IM structures are: A15 = cP8-Cr_3Si; B1 = cF8-NaCl; B2 = cP2-ClCs; B27 = oP8-FeB-b; C14 = hP12-MgZn2; C15 = cF24-Cu_2Mg; C15b = cF24-Be_5Au; C16 = tI12-Al_2Cu; C32 = hP3-AlB_2; C36 = hP24-$MgNi_2$; Cb = oF48-Mg_2Cu, $D0_{11}$ = oP16-Fe_3C, $D0_{22}$ = tI8-Al_3Ti; $D8_2$ = cI52-Cu_5Zn_8; $D8_4$ = cF116-$Cr_{23}C_6$; $D8_8$ = hP16-Mn_5Si_3; $D8_e$ = cI162-$Mg_{32}(Zn_2Al)_{49}$; D8I = tI325-Cr_5B_3; η = hP16-$TiNi_3$; $L2_1$ = cF16-$AlCu_2Mn$, χ = cI58-Ti_3Re_{24}; $L1_0$ = tP2-AuCu; $L1_2$ = cP24-$AuCu_3$; μ = hR13-Fe_7W_6; R = hR21-$TiNi_2$; σ = tP30-CrFe. Some reported compounds have two different structures, both structures are listed and separated with a slash. For such compounds, "phase name" will be left empty. In addition, structure prototypes of certain compounds are yet unknown. This is abbreviated as Unkn.

The 142 IM-containing alloys use a total of 25 elements. This includes 1 alkali metal (Li); 1 alkaline earth metal (Mg); 18 transition metals (Ag, Co, Cr, Cu, Fe, Hf, Mn, Mo, Nb, Ni, Pt, Ta, Ti, V, W, Y, Zn, Zr); 2 basic metals (Al, Sn); 2 metalloids (B, Si) and 1 non-metal (C). To understand the prevalence of these elements in our alloy database, the relative frequency of usage of each element is calculated and plotted in Figure 1. Co, Cr, Fe, and Ni are the most common elements in this study and have relative frequencies higher than 66%. Al (56%) and Ti (49%) also have very high relative frequencies. Three other elements, including Cu (28%), V (25%), and Zr (21%), have relative frequencies higher than 20%. All other elements have relative frequencies below 20%.

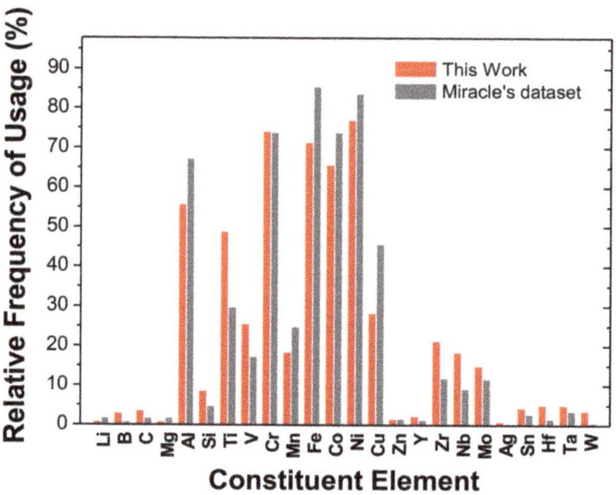

Figure 1. Relative frequencies of usage of various elements in the present study and in Miracle and Senkov's work, data from [3].

Miracle and Senkov published a comprehensive review on HEAs very recently [3]. They constructed an HEA database with a total of 408 unique alloys, regardless of the phase type present in the alloy. Thus, their alloys can serve as a good comparison group to ours. The frequency with which each element is used in their database has been calculated [3]. This data is converted to relative frequency and plotted in Figure 1. As seen in Figure 1, Co, Cr, Fe, and Ni still have the highest relative frequencies (74% or higher). Other elements having relative frequencies higher than 20% include Al (67%), Cu (46%), Ti (30%), and Mn (25%). Therefore, the seven most common elements in both databases are the same, but the relative sequence is slightly different. The popularity of these elements in both databases is largely because most of them belong to the two classic HEA systems [2]: Al-Co-Cu-Cr-Fe-Ni and Al-Co-Cr-Fe-Ni. These systems and their derivative systems (systems having two or fewer different principal elements) constitute the largest HEA family [2,3]. Thus, the popularity of these elements is not unexpected.

Comparing the relative frequencies of the same element in the two databases also helps to reveal the differences between the two databases. Based on Figure 1, only six elements have lower relative frequencies in our database than in Miracle and Senkov's, namely Al, Fe, Co, Ni, Cu, and Mn. Other elements all have higher relative frequencies of usage in our database than in Miracle's. The reason for the former is two-fold. Firstly, in our database, alloys with similar compositions and identical phase constituents are only listed once (see Section 2). Many alloys containing Al, Fe, Co, Ni, Cu, and Mn are thus excluded, leading to lower relative frequencies. For example, Miracle and Senkov's database lists 29 IM-containing alloys in the Al-Co-Cr-Cu-Fe-Ni system, but ours lists only two. The second is that many simple solid solution alloys are also composed of these elements. They are also excluded in the present study. The reason for the latter is evident—the majority of elements typically lead to the formation of IM phases. Thus, in a database of IM-containing alloys, they exhibit higher relative frequencies.

The alloys in our database can be roughly categorized into three groups. The first is the CoCrFeNi-based alloys. These are defined here as alloys containing more than 60 at.% of Co, Cr, Fe, and Ni. The second is the refractory alloys, defined here as alloys containing more that 60 at.% of refractory elements (Cr, Hf, Mo, Nb, Ta, Ti, V, W, and Zr). The third is other alloys, which includes all alloys that do not belong to the previous groups. In our database, there are 86 CoCrFeNi-based alloys (60.6%), 25 refractory alloys (17.6%), and 31 other alloys (21.8%).

It is also worth noting that processing route can have an evident effect on the type of IM phase present in the alloys. For example, the IM phases in arc-melted [42] and cladded $AlCoCrFeNiTi_{1.5}$ [43] alloys are different. Another example is the $Al_{0.3}CoCrFeNi$ alloy [44], in which different annealing temperature can lead to completely different IM phase type, and thus different mechanical properties.

3.2. The Most Prevalent IM Structures

Figure 2a shows the relative frequencies of occurrence for the most common IM structures in the present study. The sequence, in order of decreasing prevalence, is (values in parentheses indicate number of occurrence): Laves (44), σ (38), B2 (37), $L1_2$ (15), $L2_1$ (8), μ (5), Ti_2Ni (5), $D8_4$ (4), χ (4), η (4), and $D8_8$ (4). The Laves phase includes three closely related structure types: C14, C15 and C36. Their respective values are indicated in the corresponding bars. From Figure 2a, Laves, σ, B2, $L1_2$ and $L2_1$ are the five most prevalent structures, with Laves, σ, and B2 appearing in more than 25% of the alloys. $L1_2$ and $L2_1$, on the other hand, appear in approximately 5–10% of the alloys. All other structures have relative frequencies lower than 5%. In other words, the popularity of the structures is quite non-uniform. This is not unexpected because among all identified IM structure types, the distribution of popularity is also quite non-uniform. For example, about 50% of all known IM compounds belong to 44 most popular structure types [38]—that is less than 2% of the known structure types! Additionally, as shown in Figure 1, the usage of the elements in our HEAs is also quite non-uniform. This will further concentrate the structure of the IM phases to certain structure types because of compositional similarity.

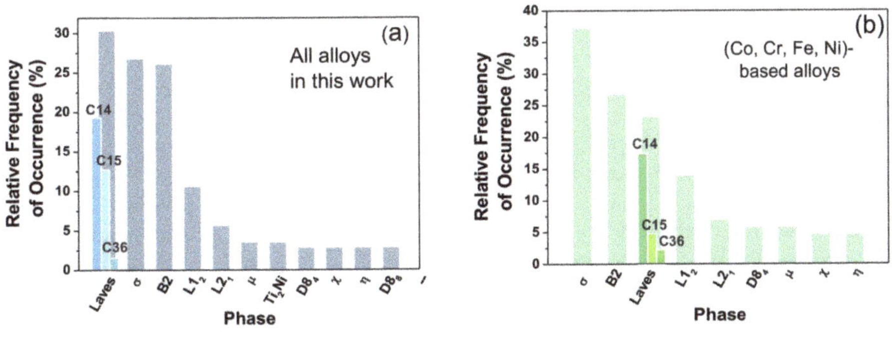

Figure 2. Relative frequencies of occurrence for the most common IM structures in this study: (**a**) Based on all alloys in the database; (**b**) Based on (Co, Cr, Fe, Ni)-based alloys only.

As mentioned previously, it is meaningful to compare the results of the present database and Miracle and Senkov's. Therefore, the relative frequencies of common IM structures in their alloy database are also calculated and plotted in Figure 3a. The most common IM structures in order of decreasing prevalence are (values in parentheses indicate number of occurrence): B2 (177), σ (60), Laves (58) [C14 (50), C15 (8)], $L1_2$ (15), $E9_3$ (6), Ti_2Ni (5), $D0_2$ (4), and $L2_1$ (4). Thus, the top four most common structures in the two databases are the same, but the relative sequence is different. This difference largely originates from the exclusion of compositionally similar alloys in the present study. For examples, there are a total of 334 IM-containing microstructure reports in Miracle and Senkov's study. However, the Al-Co-Cr-Fe-Ni alloy system alone accounts for 86 (or 25.7%) of them! This is because Al-Co-Cr-Fe-Ni is one of the two "classic" systems that have been studied most thoroughly. Thus, the effect of composition has been probed in considerable detail. For instance, Al_xCoCrFeNi alloys with x = 0.3, 0.44, 0.7, 0.75, 0.8, 0.876, 0.9, 1, 1.2, 1.25, 1.5, 1.8, 2, 2.08, 2.5 and 3 have all been reported. Besides composition, the effect of processing route is also probed, leading to even more microstructure reports. Unfortunately, the IM phases in the Al-Co-Cr-Fe-Ni system are very similar—46 of them contain B2. Due to these reasons, the frequency of the B2 phase is evidently overestimated. This probably explains why B2 tops the list in Miracle and Senkov's database. Although their work is comprehensive and of great value, we believe our analysis better represents the overall reality by reducing the bias from similar alloys.

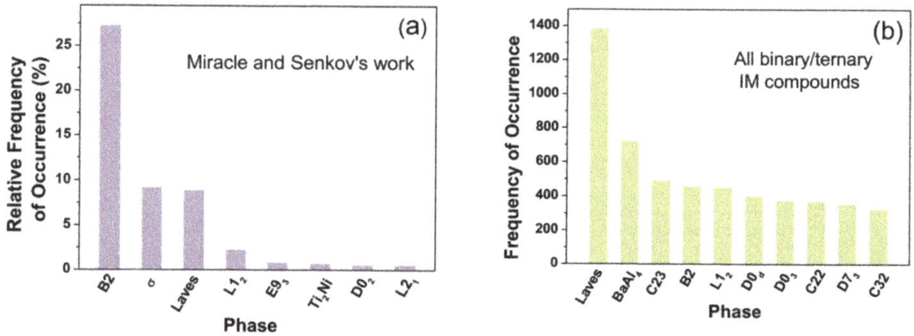

Figure 3. (**a**) Relative frequencies of occurrence for the most common IM structures in Miracle and Senkov's work, data from [3]; (**b**) Frequency of occurrence for the most common IM structures in all binary/ternary IM phases, data from [38].

Ferro and Saccone [38] calculated the most common structure types based on Villars and Calver's collection of about 26,000 known IM phases [36]. It is meaningful to compare our results with this over trend in all IM phases. Figure 3b shows the most common structure types among all binary and ternary IM phases [38]. Please note that the NaCl structure is omitted in Figure 3b because it is typically associated with non-metallic elements such as O and N (i.e., mostly ceramic phases). The five most common structures in order of decreasing prevalence are: Laves (including both C14 and C15 structures), $BaAl_4$, C23, B2, and $L1_2$. Compared with the top-five list in our database, it is seen that the Laves, B2, and $L1_2$ phases remain popular in HEAs, but the $BaAl_4$ and Co_2Si structures do not appear in our HEA database at all. This distinction is not difficult to understand if we take a closer look at compounds with these structures. $BaAl_4$-type compounds almost always contain elements in the lanthanide and actinide series, which are still very rarely used in HEAs [36]. The case for the C23 structure is similar—C23-type compounds are typically formed in compounds containing lanthanide, IIA, and IVA elements [36]. As will be shown later, IM structures formed in HEAs are always structures that existing in their subsystems. Consequently, if the necessary component of a particular structure is absent in an alloy, it is unlikely that such structure will form. By and large, the differences between the common IM structures in the present study and that in all binary/ternary IM compounds is probably because the explored compositional space in HEAs is still extremely limited.

For the (Co, Cr, Fe, Ni)-based alloys, the five most common phases (Figure 2b) listed in order of decreasing prevalence are: σ (32), B2 (23), Laves (20), $L1_2$ (12), and $L2_1$ (6). The common phases are same as that in all alloys, but the relative popularity is different. This again suggests that the IM phases present are still dependent on the composition. As will be shown later, the IM phases in HEAs are basically the IM phases that already exist in its binary and/or ternary subsystems. Thus, the choice of element is directly related to the type of IM phase.

3.3. Structure Inheritance in HEAs

Before going further, it is useful here to explain the term "structural inheritance" using ternary systems as examples. IM phases in ternary phase diagrams are very often extended from the boundary of the diagrams. In other words, they are often solid solutions of binary IM phases. Thus, the IM structure already exists in at least one of the three binary subsystems. In this work, this is described as the inheritance of the IM structure in the ternary alloy from its binary subsystems. In some cases, the IM phase region can even connect from one edge of the triangle to another, as illustrated in the imaginary phase diagram in Figure 4a. Similarly, in a quaternary system, the IM structures can be inherited from binary and ternary subsystems.

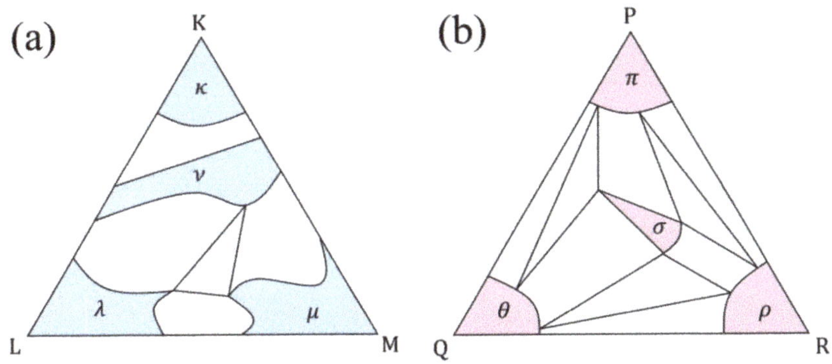

Figure 4. Imaginary phase diagrams illustrating the inheritance of IM phases in ternary systems. Colored areas represent single phase regions. (**a**) A phase inherited from two binary subsystems (the ν phase); (**b**) A phase that is original to the ternary systems (the σ phase).

However, there are cases where the IM phase cannot be found in any of the binary subsystems. Therefore, the IM phase is not inherited and is original to the ternary alloy. In such cases, the phase region of the IM phase is in the interior of the phase diagram, as illustrated in Figure 4b. Similar situations can be seen in quaternary systems, too. For example, in the Al-Co-Fe-Ti system, the TiFeCoAl phase (cF16-LiMgPdSn) is original and does not exist in any of its binary or ternary subsystems. What about the case in HEAs? Are the IM structures in HEAs inherited from their subsystems? Do original structures exist? If so, what are they and how often do they appear?

Table 1 shows the IM structures in each HEA collected in this study. We performed a detailed analysis on these structures to identify whether or not the structure is inherited. In some trivial cases, this can be confirmed directly from the chemical formula. For example, a Fe_2Ti phase in the CoCrCuFeNiTi alloy is clearly inherited from the Fe-Ti binary system. In other cases, this is done by comparing the structure prototypes of the IM phases in an HEA and those in its subsystems. The result is indicated in the "IM Structure Found in" column in Table 1. As shown in the table, all the IM structures in the 142 alloys can be found in the subsystems of the respective alloy. In fact, in more than 90% of the alloys (129 out of 142 alloys), the IM structures are found in binary subsystems. In the remaining 13 alloys, the structures cannot be found in binary subsystems and are seen only in ternary ones. A typical reason for this is that the structure is intrinsically ternary (i.e., three different elements are required to form that structure). This is typically reflected by a ternary prototype. The best example is the $L2_1$ or Heusler phase (cF16-$AlCu_2Mn$). 8 of the 13 remaining alloys contain the $L2_1$ phase. There are also structures that are not intrinsically ternary, but are still found in ternary subsystems only. The most notable case is the χ phase (found in the Cr-Fe-Ti system).

Based on the above, for all the HEAs in this study, the IM structures are inherited from their respective binary/ternary subsystems. This information has important implications. To predict the IM phases in HEAs, whether via empirical or CALPHAD approaches, a list of candidate phases is necessary so that the relative stabilities of the candidate phases can be compared to determine the stable phase. If the IM structures in an HEA are frequently irrelevant to its binary and ternary subsystems, given that phase diagrams of quaternary and higher-order systems are very rare, the prediction of IM phases in HEAs would be extremely difficult. Our results; however, show that the candidate IM phases basically resides in the binary and ternary subsystems, so phase prediction is feasible. Another way of looking at this is that if the composing elements in an HEA do not form a particular IM structure, it is unlikely that that phase will appear in the alloy.

Our results also suggest that when considering the candidate IM phases in an HEA, it is reasonable to consider just the phases in binary subsystems, plus certain important structures from known ternary combinations. This is because in more than 90% of the alloys, the IM phases exist in their binary subsystems. Moreover, those existing only in ternary subsystems are highly focused on certain element combinations. For example, the $L2_1$ phase is related to the Al-Ni-Ti combination, and the χ phase is related to the Cr-Fe-Ti combination. Further analysis of the dependence of IM phase type on composition is underway and will be published later.

4. Conclusions

A database of IM-containing HEAs has been constructed by examining related publications. The database consists of 142 unique alloys. Care is taken to avoid compositional repetition so that the diversity in the database is enhanced and statistical bias is reduced. The usage of elements in these alloys is very non-uniform and highly concentrated to Co, Cr, Fe, Ni, Al, and Ti, all appearing in 50% or more of the alloys.

Based on statistical analysis of the database, it is found that the five most prevalent IM structures in all HEAs are Laves, σ, B2, $L1_2$, and $L2_1$. Those in (Co, Cr, Fe, Ni)-based HEAs are the same, but with a different sequence. These results are evidently different from the overall trend in known binary/ternary IM structures mainly because the non-uniform usage of the elements in the alloys, which prevents the formation of many structures.

All the IM structures in the HEAs surveyed here are existing structures in the binary/ternary subsystems of the respective alloys. This means that HEAs tend to inherit IM structures from their subsystems, instead of forming new structures irrelevant to their subsystems. It seems that the compositional complexity in HEAs does not trigger additional structural complexity. Therefore, when predicting IM phases in HEAs, it is probably reasonable to consider only the IM phase types existing in the binary and ternary subsystems of the alloys, and ignore irrelevant phases. These findings have important implications in the future design and development of HEAs.

Author Contributions: M.H.T. designed the research and wrote the paper. R.C.T and M.H.T. constructed the database. R.C.T, M.H.T., T.C., and W.F.H analyzed the data and did the illustrations.

Funding: This research was financially supported by the Ministry of Science and Technology, Taiwan (under grant MOST 107-2218-E-005-006) and the High Entropy Materials Center from the Featured Areas Research Center Program within the framework of the Higher Education Sprout Project by the Ministry of Education (MOE), Taiwan and from project MOST 107-3017-F-007-003 by Ministry of Science and Technology (MOST), Taiwan.

Conflicts of Interest: The authors declare no conflict of interest.

References

1. Yeh, J.W.; Chen, S.K.; Lin, S.J.; Gan, J.Y.; Chin, T.S.; Shun, T.T.; Tsau, C.H.; Chang, S.Y. Nanostructured high-entropy alloys with multiple principal elements: Novel alloy design concepts and outcomes. *Adv. Eng. Mater.* **2004**, *6*, 299–303. [CrossRef]
2. Tsai, M.H.; Yeh, J.W. High-Entropy Alloys: A Critical Review. *Mater. Res. Lett.* **2014**, *2*, 107–123. [CrossRef]
3. Miracle, D.B.; Senkov, O.N. A critical review of high entropy alloys and related concepts. *Acta Mater.* **2017**, *122*, 448–511. [CrossRef]
4. Zhang, Y.; Zhou, Y.J.; Lin, J.P.; Chen, G.L.; Liaw, P.K. Solid-solution phase formation rules for multi-component alloys. *Adv. Eng. Mater.* **2008**, *10*, 534–538. [CrossRef]

5. Guo, S.; Liu, C.T. Phase stability in high entropy alloys: Formation of solid-solution phase or amorphous phase. *Prog. Nat. Sci. Mater. Int.* **2011**, *21*, 433–446. [CrossRef]
6. Guo, S. Phase selection rules for cast high entropy alloys: An overview. *Mater. Sci. Technol.* **2015**, *31*, 1223–1230. [CrossRef]
7. Tsai, K.Y.; Tsai, M.H.; Yeh, J.W. Sluggish diffusion in Co–Cr–Fe–Mn–Ni high-entropy alloys. *Acta Mater.* **2013**, *61*, 4887–4897. [CrossRef]
8. Zaddach, A.J.; Niu, C.; Koch, C.C.; Irving, D.L. Mechanical Properties and Stacking Fault Energies of NiFeCrCoMn High-Entropy Alloy. *JOM* **2013**, *65*, 1780–1789. [CrossRef]
9. Liu, W.H.; Wu, Y.; He, J.Y.; Nieh, T.G.; Lu, Z.P. Grain growth and the Hall-Petch relationship in a high-entropy FeCrNiCoMn alloy. *Scr. Mater.* **2013**, *68*, 526–529. [CrossRef]
10. Laplanche, G.; Gadaud, P.; Horst, O.; Otto, F.; Eggeler, G.; George, E.P. Temperature dependencies of the elastic moduli and thermal expansion coefficient of an equiatomic, single-phase CoCrFeMnNi high-entropy alloy. *J. Alloys Compd.* **2015**, *623*, 348–353. [CrossRef]
11. Laplanche, G.; Bonneville, J.; Varvenne, C.; Curtin, W.A.; George, E.P. Thermal activation parameters of plastic flow reveal deformation mechanisms in the CrMnFeCoNi high-entropy alloy. *Acta Mater.* **2018**, *143*, 257–264. [CrossRef]
12. Gludovatz, B.; Hohenwarter, A.; Catoor, D.; Chang, E.H.; George, E.P.; Ritchie, R.O. A fracture-resistant high-entropy alloy for cryogenic applications. *Science* **2014**, *345*, 1153–1158. [CrossRef] [PubMed]
13. Zhang, Y.; Stocks, G.M.; Jin, K.; Lu, C.; Bei, H.; Sales, B.C.; Wang, L.; Beland, L.K.; Stoller, R.E.; Samolyuk, G.D.; et al. Influence of chemical disorder on energy dissipation and defect evolution in concentrated solid solution alloys. *Nat. Commun.* **2015**, *6*, 8376. [CrossRef] [PubMed]
14. Zhang, Z.; Mao, M.M.; Wang, J.; Gludovatz, B.; Zhang, Z.; Mao, S.X.; George, E.P.; Yu, Q.; Ritchie, R.O. Nanoscale origins of the damage tolerance of the high-entropy alloy CrMnFeCoNi. *Nat. Commun.* **2015**, *6*, 10143. [CrossRef] [PubMed]
15. Zou, Y.; Ma, H.; Spolenak, R. Ultrastrong ductile and stable high-entropy alloys at small scales. *Nat. Commun.* **2015**, *6*, 7748. [CrossRef] [PubMed]
16. Li, Z.; Pradeep, K.G.; Deng, Y.; Raabe, D.; Tasan, C.C. Metastable high-entropy dual-phase alloys overcome the strength–ductility trade-off. *Nature* **2016**, *534*, 227–230. [CrossRef] [PubMed]
17. Gali, A.; George, E.P. Tensile properties of high- and medium-entropy alloys. *Intermetallics* **2013**, *39*, 74–78. [CrossRef]
18. Gludovatz, B.; George, E.P.; Ritchie, R.O. Processing, Microstructure and Mechanical Properties of the CrMnFeCoNi High-Entropy Alloy. *JOM* **2015**, *67*, 2262–2270. [CrossRef]
19. Miracle, D.; Miller, J.; Senkov, O.; Woodward, C.; Uchic, M.; Tiley, J. Exploration and Development of High Entropy Alloys for Structural Applications. *Entropy* **2014**, *16*, 494–525. [CrossRef]
20. Lu, Z.P.; Wang, H.; Chen, M.W.; Baker, I.; Yeh, J.W.; Liu, C.T.; Nieh, T.G. An assessment on the future development of high-entropy alloys: Summary from a recent workshop. *Intermetallics* **2015**, *66*, 67–76. [CrossRef]
21. Tsai, M.H. Three Strategies for the Design of Advanced High-Entropy Alloys. *Entropy* **2016**, *18*, 252. [CrossRef]
22. Chuang, M.H.; Tsai, M.H.; Wang, W.R.; Lin, S.J.; Yeh, J.W. Microstructure and wear behavior of $Al_xCo_{1.5}CrFeNi_{1.5}Ti_y$ high-entropy alloys. *Acta Mater.* **2011**, *59*, 6308–6317. [CrossRef]
23. Chuang, M.H.; Tsai, M.H.; Tsai, C.W.; Yang, N.H.; Chang, S.Y.; Yeh, J.W.; Chen, S.K.; Lin, S.J. Intrinsic surface hardening and precipitation kinetics of $Al_{0.3}CrFe_{1.5}MnNi_{0.5}$ multi-component alloy. *J. Alloys Compd.* **2013**, *551*, 12–18. [CrossRef]
24. He, J.Y.; Wang, H.; Huang, H.L.; Xu, X.D.; Chen, M.W.; Wu, Y.; Liu, X.J.; Nieh, T.G.; An, K.; Lu, Z.P. A precipitation-hardened high-entropy alloy with outstanding tensile properties. *Acta Mater.* **2016**, *102*, 187–196. [CrossRef]
25. Liu, W.H.; Lu, Z.P.; He, J.Y.; Luan, J.H.; Wang, Z.J.; Liu, B.; Liu, Y.; Chen, M.W.; Liu, C.T. Ductile $CoCrFeNiMo_x$ high entropy alloys strengthened by hard intermetallic phases. *Acta Mater.* **2016**, *116*, 332–342. [CrossRef]
26. Wani, I.S.; Bhattacharjee, T.; Sheikh, S.; Lu, Y.P.; Chatterjee, S.; Bhattacharjee, P.P.; Guo, S.; Tsuji, N. Ultrafine-grained $AlCoCrFeNi_{2.1}$ eutectic high-entropy alloy. *Mater. Res. Lett.* **2016**, *4*, 174–179. [CrossRef]
27. Ming, K.S.; Bi, X.F.; Wang, J. Precipitation strengthening of ductile $Cr_{15}Fe_{20}Co_{35}Ni_{20}Mo_{10}$ alloys. *Scr. Mater.* **2017**, *137*, 88–93. [CrossRef]

28. Troparevsky, M.C.; Morris, J.R.; Kent, P.R.C.; Lupini, A.R.; Stocks, G.M. Criteria for Predicting the Formation of Single-Phase High-Entropy Alloys. *Phys. Rev. X* **2015**, *5*, 011041. [CrossRef]
29. Senkov, O.N.; Miracle, D.B. A new thermodynamic parameter to predict formation of solid solution or intermetallic phases in high entropy alloys. *J. Alloys Compd.* **2016**, *658*, 603–607. [CrossRef]
30. Ye, Y.F.; Wang, Q.; Lu, J.; Liu, C.T.; Yang, Y. Design of high entropy alloys: A single-parameter thermodynamic rule. *Scr. Mater.* **2015**, *104*, 53–55. [CrossRef]
31. Tsai, M.H.; Li, J.H.; Fan, A.C.; Tsai, P.H. Incorrect predictions of simple solid solution high entropy alloys: Cause and possible solution. *Scr. Mater.* **2017**, *127*, 6–9. [CrossRef]
32. Tsai, M.H.; Fan, A.C.; Wang, H.A. Effect of atomic size difference on the type of major intermetallic phase in arc-melted CoCrFeNiX high-entropy alloys. *J. Alloys Compd.* **2017**, *695*, 1479–1487. [CrossRef]
33. Tsai, M.H.; Tsai, K.Y.; Tsai, C.W.; Lee, C.; Juan, C.C.; Yeh, J.W. Criterion for sigma phase formation in Cr- and V-containing high-entropy alloys. *Mater. Res. Lett.* **2013**, *1*, 207–212. [CrossRef]
34. Tsai, M.H.; Chang, K.C.; Li, J.H.; Tsai, R.C.; Cheng, A.H. A second criterion for sigma phase formation in high-entropy alloys. *Mater. Res. Lett.* **2016**, *4*, 90–95. [CrossRef]
35. Yurchenko, N.; Stepanov, N.; Salishchev, G. Laves-phase formation criterion for high-entropy alloys. *J. Energy Mater.* **2016**, *11*, 17–22. [CrossRef]
36. Villars, P.; Calvert, L.D. (Eds.) *Pearson's Handbook of Crystallographic Data for Intermetallic Phases*, 2nd ed.; ASM International: Materials Park, OH, USA, 1991.
37. He, J.Y.; Liu, W.H.; Wang, H.; Wu, Y.; Liu, X.J.; Nieh, T.G.; Lu, Z.P. Effects of Al addition on structural evolution and tensile properties of the FeCoNiCrMn high-entropy alloy system. *Acta Mater.* **2014**, *62*, 105–113. [CrossRef]
38. Ferro, P.; Saccone, A. Structure of Intermetallic Compounds and Phases. In *Physical Metallurgy*, 4th ed.; Cahn, R.W., Haasen, P., Eds.; Elsevier Science B.V.: Amsterdam, The Netherlands, 1996; Volume 1, pp. 206–369.
39. Villars, P.; Prince, A.; Okamoto, H. *Handbook of Ternary Alloy Phase Diagrams*; ASM International: Materials Park, OH, USA, 1995; Volume 3.
40. Baker, H. (Ed.) *ASM Handbook: Alloy Phase Diagrams*; ASM International: Materials Park, OH, USA, 1992.
41. Villars, P.; Cenzual, K.; Gladyshevskii, R. *Handbook of Inorganic Substances 2015*; De Gruyter: Berlin, Germany, 2015.
42. Zhou, Y.; Zhang, Y.; Wang, Y.; Chen, G. Solid solution alloys of $AlCoCrFeNiTi_x$ with excellent room-temperature mechanical properties. *Appl. Phys. Lett.* **2007**, *90*, 181904. [CrossRef]
43. Wu, C.; Zhang, S.; Zhang, C.; Zhang, H.; Dong, S. Phase evolution and cavitation erosion-corrosion behavior of $FeCoCrAlNiTi_x$ high entropy alloy coatings on 304 stainless steel by laser surface alloying. *J. Alloys Compd.* **2017**, *698*, 761–770. [CrossRef]
44. Gwalani, B.; Soni, V.; Choudhuri, D.; Lee, M.; Hwang, J.; Nam, S.; Ryu, H.; Hong, S.; Banerjee, R. Stability of ordered $L1_2$ and B2 precipitates in face centered cubic based high entropy alloys-Al0.3CoFeCrNi and Al0.3CuFeCrNi2. *Scr. Mater.* **2016**, *123*, 130–134. [CrossRef]
45. Hsu, U.S.; Hung, U.D.; Yeh, J.W.; Chen, S.K.; Huang, Y.S.; Yang, C.C. Alloying behavior of iron, gold and silver in AlCoCrCuNi-based equimolar high-entropy alloys. *Mater. Sci. Eng. A* **2007**, *460*, 403–408. [CrossRef]
46. Lee, C.; Chen, Y.; Hsu, C.; Yeh, J.; Shih, H. Effect of Boron on the Corrosion Properties of Al0.5CoCrCuFeNiBx High Entropy Alloys in 1N Sulfuric Acid. In Proceedings of the 209th Electrochemical Society (ECS) Meeting, Denver, CO, USA, 7–12 May 2006.
47. Zhang, H.; He, Y.; Pan, Y. Enhanced hardness and fracture toughness of the laser-solidified FeCoNiCrCuTiMoAlSiB0.5 high-entropy alloy by martensite strengthening. *Scr. Mater.* **2013**, *69*, 342–345. [CrossRef]
48. Shun, T.-T.; Du, Y.-C. Age hardening of the Al0.3CoCrFeNiC0.1 high entropy alloy. *J. Alloys Compd.* **2009**, *478*, 269–272. [CrossRef]
49. Fang, S.; Chen, W.; Fu, Z. Microstructure and mechanical properties of twinned Al0.5CrFeNiCo0.3C0.2 high entropy alloy processed by mechanical alloying and spark plasma sintering. *Mater. Des. (1980-2015)* **2014**, *54*, 973–979. [CrossRef]
50. Praveen, S.; Murty, B.S.; Kottada, R.S. Alloying behavior in multi-component AlCoCrCuFe and NiCoCrCuFe high entropy alloys. *Mater. Sci. Eng. A* **2012**, *534*, 83–89. [CrossRef]
51. Zhou, Y.J.; Zhang, Y.; Wang, Y.L.; Chen, G.L. Microstructure and compressive properties of multicomponent $Al_x(TiVCrMnFeCoNiCu)_{100-x}$ high-entropy alloys. *Mater. Sci. Eng. A* **2007**, *454–455*, 260–265. [CrossRef]

52. Yuhu, F.; Yunpeng, Z.; Hongyan, G.; Huimin, S.; Li, H. AlNiCrFexMo0.2CoCu High Entropy Alloys Prepared by Powder Metallurgy. *Rare Met. Mater. Eng.* **2013**, *42*, 1127–1129. [CrossRef]
53. Manzoni, A.M.; Daoud, H.M.; Voelkl, R.; Glatzel, U.; Wanderka, N. Influence of W, Mo and Ti trace elements on the phase separation in Al8Co17Cr17Cu8Fe17Ni33 based high entropy alloy. *Ultramicroscopy* **2015**, *159 Pt 2*, 265–271. [CrossRef]
54. Ng, C.; Guo, S.; Luan, J.; Shi, S.; Liu, C.T. Entropy-driven phase stability and slow diffusion kinetics in an $Al_{0.5}$CoCrCuFeNi high entropy alloy. *Intermetallics* **2012**, *31*, 165–172. [CrossRef]
55. Singh, S.; Wanderka, N.; Murty, B.S.; Glatzel, U.; Banhart, J. Decomposition in multi-component AlCoCrCuFeNi high-entropy alloy. *Acta Mater.* **2011**, *59*, 182–190. [CrossRef]
56. Jensen, J.; Welk, B.; Williams, R.; Sosa, J.; Huber, D.; Senkov, O.; Viswanathan, G.; Fraser, H. Characterization of the microstructure of the compositionally complex alloy Al1Mo0.5Nb1Ta0.5Ti1Zr1. *Scr. Mater.* **2016**, *121*, 1–4. [CrossRef]
57. Chen, M.-R.; Lin, S.-J.; Yeh, J.-W.; Chen, S.-K.; Huang, Y.-S.; Tu, C.-P. Microstructure and Properties of $Al_{0.5}$CoCrCuFeNiTi_x (x = 0-2.0) High-Entropy Alloys. *Jpn. Inst. Met.* **2006**, *47*, 1395–1401.
58. Chen, M.-R.; Lin, S.-J.; Yeh, J.-W.; Chuang, M.-H.; Chen, S.-K.; Huang, Y.-S. Effect of vanadium addition on the microstructure, hardness, and wear resistance of Al0.5CoCrCuFeNi high-entropy alloy. *Metall. Mater. Trans. A-Phys. Metall. Mater. Sci.* **2006**, *37*, 1363–1369. [CrossRef]
59. Young, K.; Ouchi, T.; Huang, B.; Reichman, B.; Fetcenko, M.A. Studies of copper as a modifier in C14-predominant AB2 metal hydride alloys. *J. Power Sources* **2012**, *204*, 205–212. [CrossRef]
60. Wang, Z.; Wang, X.; Yue, H.; Shi, G.; Wang, S. Microstructure, thermodynamics and compressive properties of AlCoCrCuMn-x (x=Fe, Ti) high-entropy alloys. *Mater. Sci. Eng. A* **2015**, *627*, 391–398. [CrossRef]
61. Yeh, J.W.; Lin, S.J.; Chin, T.S.; Gan, J.Y.; Chen, S.K.; Shun, T.T.; Tsau, C.H.; Chou, S.Y. Formation of simple crystal structures in Cu-Co-Ni-Cr-Al-Fe-Ti-V alloys with multiprincipal metallic elements. *Metall. Mater. Trans. A* **2004**, *35*, 2533–2536. [CrossRef]
62. Hu, Z.; Zhan, Y.; Zhang, G.; She, J.; Li, C. Effect of rare earth Y addition on the microstructure and mechanical properties of high entropy AlCoCrCuNiTi alloys. *Mater. Des.* **2010**, *31*, 1599–1602. [CrossRef]
63. Juan, C.C.; Hsu, C.Y.; Tsai, C.W.; Wang, W.R.; Sheu, T.S.; Yeh, J.W.; Chen, S.K. On microstructure and mechanical performance of AlCoCrFeMo0.5Nix high-entropy alloys. *Intermetallics* **2013**, *32*, 401–407. [CrossRef]
64. Ma, S.; Zhang, Y. Effect of Nb addition on the microstructure and properties of AlCoCrFeNi high-entropy alloy. *Mater. Sci. Eng. A* **2012**, *532*, 480–486. [CrossRef]
65. Zhang, H.; He, Y.-Z.; Pan, Y.; Pei, L.-Z. Phase selection, microstructure and properties of laser rapidly solidified FeCoNiCrAl2Si coating. *Intermetallics* **2011**, *19*, 1130–1135. [CrossRef]
66. Wang, L.; Chen, C.; Yeh, J.; Ke, S. The microstructure and strengthening mechanism of thermal spray coating NixCo0.6Fe0.2CrySizAlTi0.2 high-entropy alloys. *Mater. Chem. Phys.* **2011**, *126*, 880–885. [CrossRef]
67. Lee, C.-F.; Shun, T.-T. Effect of Fe content on microstructure and mechanical properties of $Al_{0.5}$CoCrFe_xNiTi$_{0.5}$ high-entropy alloys. *Mater. Charact.* **2016**, *114*, 179–184. [CrossRef]
68. Zhang, K.; Fu, Z.; Zhang, J.; Wang, W.; Wang, H.; Wang, Y.; Zhang, Q.; Shi, J. Microstructure and mechanical properties of CoCrFeNiTiAl*x* high-entropy alloys. *Mater. Sci. Eng. A* **2009**, *508*, 214–219. [CrossRef]
69. Chen, J.; Niu, P.; Liu, Y.; Lu, Y.; Wang, X.; Peng, Y.; Liu, J. Effect of Zr content on microstructure and mechanical properties of AlCoCrFeNi high entropy alloy. *Mater. Des.* **2016**, *94*, 39–44. [CrossRef]
70. Lee, C.-F.; Shun, T.-T. Age hardening of the Al0.5CoCrNiTi0.5 high-entropy alloy. *Metall. Mater. Trans. A-Phys. Metall. Mater. Sci.* **2014**, *45*, 191–195. [CrossRef]
71. Liu, L.; He, L.; Qi, J.; Wang, B.; Zhao, Z.; Shang, J.; Zhang, Y. Effects of Sn element on microstructure and properties of SnxAl2.5FeCoNiCu multi-component alloys. *J. Alloys Compd.* **2016**, *654*, 327–332. [CrossRef]
72. Liu, L.; Zhu, J.; Hou, C.; Li, J.; Jiang, Q. Dense and smooth amorphous films of multicomponent FeCoNiCuVZrAl high-entropy alloy deposited by direct current magnetron sputtering. *Mater. Des.* **2013**, *46*, 675–679. [CrossRef]
73. Zhuang, Y.; Liu, W.; Chen, Z.; Xue, H.; He, J. Effect of elemental interaction on microstructure and mechanical properties of FeCoNiCuAl alloys. *Mater. Sci. Eng. A* **2012**, *556*, 395–399. [CrossRef]
74. Mohanty, S.; Gurao, N.; Biswas, K. Sinter ageing of equiatomic Al20Co20Cu20Zn20Ni20 high entropy alloy via mechanical alloying. *Mater. Sci. Eng. A* **2014**, *617*, 211–218. [CrossRef]

75. Hsu, C.-Y.; Juan, C.-C.; Wang, W.-R.; Sheu, T.-S.; Yeh, J.-W.; Chen, S.-K. On the superior hot hardness and softening resistance of AlCoCr$_x$FeMo0.5 Ni high-entropy alloys. *Mater. Sci. Eng. A* **2011**, *528*, 3581–3588. [CrossRef]
76. Butler, T.; Alfano, J.; Martens, R.; Weaver, M. High-temperature oxidation behavior of Al-Co-Cr-Ni-(Fe or Si) multicomponent high-entropy alloys. *JOM* **2015**, *67*, 246–259. [CrossRef]
77. Maulik, O.; Kumar, D.; Kumar, S.; Fabijanic, D.M.; Kumar, V. Structural evolution of spark plasma sintered AlFeCuCrMg*x* (x= 0, 0.5, 1, 1.7) high entropy alloys. *Intermetallics* **2016**, *77*, 46–56. [CrossRef]
78. Khanchandani, H.; Sharma, P.; Kumar, R.; Maulik, O.; Kumar, V. Effect of sintering on phase evolution in AlMgFeCuCrNi4.75 high entropy alloy. *Adv. Powder Technol.* **2016**, *27*, 289–294. [CrossRef]
79. Guo, S.; Ng, C.; Liu, C.T. Anomalous solidification microstructures in Co-free AlxCrCuFeNi2 high-entropy alloys. *J. Alloys Compd.* **2013**, *557*, 77–81. [CrossRef]
80. Choudhuri, D.; Gwalani, B.; Gorsse, S.; Mikler, C.; Ramanujan, R.; Gibson, M.; Banerjee, R. Change in the primary solidification phase from fcc to bcc-based B2 in high entropy or complex concentrated alloys. *Scr. Mater.* **2017**, *127*, 186–190. [CrossRef]
81. Pi, J.-H.; Pan, Y.; Zhang, L.; Zhang, H. Microstructure and property of AlTiCrFeNiCu high-entropy alloy. *J. Alloys Compd.* **2011**, *509*, 5641–5645. [CrossRef]
82. Baker, I.; Meng, F.; Wu, M.; Brandenberg, A. Recrystallization of a novel two-phase FeNiMnAlCr high entropy alloy. *J. Alloys Compd.* **2016**, *656*, 458–464. [CrossRef]
83. Tsai, M.-H.; Yuan, H.; Cheng, G.; Xu, W.; Jian, W.W.; Chuang, M.-H.; Juan, C.-C.; Yeh, A.-C.; Lin, S.-J.; Zhu, Y. Significant hardening due to the formation of a sigma phase matrix in a high entropy alloy. *Intermetallics* **2013**, *33*, 81–86. [CrossRef]
84. Dong, Y.; Gao, X.; Lu, Y.; Wang, T.; Li, T. A multi-component AlCrFe2Ni2 alloy with excellent mechanical properties. *Mater. Lett.* **2016**, *169*, 62–64. [CrossRef]
85. Dong, Y.; Lu, Y.; Kong, J.; Zhang, J.; Li, T. Microstructure and mechanical properties of multi-component AlCrFeNiMox high-entropy alloys. *J. Alloys Compd.* **2013**, *573*, 96–101. [CrossRef]
86. Lin, C.W.; Tsai, M.H.; Tsai, C.W.; Yeh, J.W.; Chen, S.K. Microstructure and aging behaviour of Al5Cr32Fe35Ni22Ti6 high entropy alloy. *Mater. Sci. Technol.* **2015**, *31*, 1165–1170. [CrossRef]
87. Young, K.; Regmi, R.; Lawes, G.; Ouchi, T.; Reichman, B.; Fetcenko, M.; Wu, A. Effects of aluminum substitution in C14-rich multi-component alloys for NiMH battery application. *J. Alloys Compd.* **2010**, *490*, 282–292. [CrossRef]
88. Chen, H.; Kauffmann, A.; Gorr, B.; Schliephake, D.; Seemüller, C.; Wagner, J.; Christ, H.-J.; Heilmaier, M. Microstructure and mechanical properties at elevated temperatures of a new Al-containing refractory high-entropy alloy Nb-Mo-Cr-Ti-Al. *J. Alloys Compd.* **2016**, *661*, 206–215. [CrossRef]
89. Chang, H.-W.; Huang, P.-K.; Davison, A.; Yeh, J.-W.; Tsau, C.-H.; Yang, C.-C. Nitride films deposited from an equimolar Al–Cr–Mo–Si–Ti alloy target by reactive direct current magnetron sputtering. *Thin Solid Films* **2008**, *516*, 6402–6408. [CrossRef]
90. Stepanov, N.; Yurchenko, N.Y.; Panina, E.; Tikhonovsky, M.; Zherebtsov, S. Precipitation-strengthened refractory Al0.5CrNbTi2V0.5 high entropy alloy. *Mater. Lett.* **2017**, *188*, 162–164. [CrossRef]
91. Yurchenko, N.Y.; Stepanov, N.; Shaysultanov, D.; Tikhonovsky, M.; Salishchev, G. Effect of Al content on structure and mechanical properties of the Al*x*CrNbTiVZr (x= 0; 0.25; 0.5; 1) high-entropy alloys. *Mater. Charact.* **2016**, *121*, 125–134. [CrossRef]
92. Huang, C.; Zhang, Y.; Shen, J.; Vilar, R. Thermal stability and oxidation resistance of laser clad TiVCrAlSi high entropy alloy coatings on Ti–6Al–4V alloy. *Surf. Coat. Technol.* **2011**, *206*, 1389–1395. [CrossRef]
93. Yang, X.; Chen, S.; Cotton, J.; Zhang, Y. Phase stability of low-density, multiprincipal component alloys containing aluminum, magnesium, and lithium. *JOM* **2014**, *66*, 2009–2020. [CrossRef]
94. Zhang, Z.; Axinte, E.; Ge, W.; Shang, C.; Wang, Y. Microstructure, mechanical properties and corrosion resistance of CuZrY/Al, Ti, Hf series high-entropy alloys. *Mater. Des.* **2016**, *108*, 106–113. [CrossRef]
95. Wang, Z.; Wu, M.; Cai, Z.; Chen, S.; Baker, I. Effect of Ti content on the microstructure and mechanical behavior of $(Fe_{36}Ni_{18}Mn_{33}Al_{13})_{100-x}Ti_x$ high entropy alloys. *Intermetallics* **2016**, *75*, 79–87. [CrossRef]
96. Poletti, M.; Branz, S.; Fiore, G.; Szost, B.; Crichton, W.; Battezzati, L. Equilibrium high entropy phases in X-NbTaTiZr (X= Al, V, Cr and Sn) multiprincipal component alloys. *J. Alloys Compd.* **2016**, *655*, 138–146. [CrossRef]

97. Tan, X.-R.; Zhang, G.-P.; Zhi, Q.; Liu, Z.-X. Effects of milling on the microstructure and hardness of Al2NbTi3V2Zr high-entropy alloy. *Mater. Des.* **2016**, *109*, 27–36. [CrossRef]
98. Yurchenko, N.Y.; Stepanov, N.D.; Tikhonovsky, M.A.; Salishchev, G.A. Phase Evolution of the AlxNbTiVZr (x= 0; 0.5; 1; 1.5) High Entropy Alloys. *Metals* **2016**, *6*, 298. [CrossRef]
99. Ding, J.; Inoue, A.; Han, Y.; Kong, F.; Zhu, S.; Wang, Z.; Shalaan, E.; Al-Marzouki, F. High entropy effect on structure and properties of (Fe, Co, Ni, Cr)-B amorphous alloys. *J. Alloys Compd.* **2017**, *696*, 345–352. [CrossRef]
100. Stepanov, N.; Yurchenko, N.Y.; Tikhonovsky, M.; Salishchev, G. Effect of carbon content and annealing on structure and hardness of the CoCrFeNiMn-based high entropy alloys. *J. Alloys Compd.* **2016**, *687*, 59–71. [CrossRef]
101. Guo, N.; Wang, L.; Luo, L.; Li, X.; Chen, R.; Su, Y.; Guo, J.; Fu, H. Microstructure and mechanical properties of in-situ MC-carbide particulates-reinforced refractory high-entropy Mo0.5NbHf0.5ZrTi matrix alloy composite. *Intermetallics* **2016**, *69*, 74–77. [CrossRef]
102. Wang, X.; Zhang, Y.; Qiao, Y.; Chen, G. Novel microstructure and properties of multicomponent $CoCrCuFeNiTi_x$ alloys. *Intermetallics* **2007**, *15*, 357–362. [CrossRef]
103. Guo, S.; Hu, Q.; Ng, C.; Liu, C. More than entropy in high-entropy alloys: Forming solid solutions or amorphous phase. *Intermetallics* **2013**, *41*, 96–103. [CrossRef]
104. Huo, W.-Y.; Shi, H.-F.; Ren, X.; Zhang, J.-Y. Microstructure and wear behavior of CoCrFeMnNbNi high-entropy alloy coating by TIG cladding. *Adv. Mater. Sci. Eng.* **2015**, *2015*, 647351. [CrossRef]
105. Stepanov, N.; Shaysultanov, D.; Salishchev, G.; Tikhonovsky, M.; Oleynik, E.; Tortika, A.; Senkov, O. Effect of V content on microstructure and mechanical properties of the $CoCrFeMnNiV_x$ high entropy alloys. *J. Alloys Compd.* **2015**, *628*, 170–185. [CrossRef]
106. Shun, T.-T.; Chang, L.-Y.; Shiu, M.-H. Microstructure and mechanical properties of multiprincipal component $CoCrFeNiMo_x$ alloys. *Mater. Charact.* **2012**, *70*, 63–67. [CrossRef]
107. Shun, T.T.; Chang, L.Y.; Shiu, M.H. Age-hardening of the CoCrFeNiMo0.85 high-entropy alloy. *Mater. Charact.* **2013**, *81*, 92–96. [CrossRef]
108. Chou, Y.; Yeh, J.; Shih, H. The effect of molybdenum on the corrosion behaviour of the high-entropy alloys Co1.5CrFeNi1.5Ti0.5Mox in aqueous environments. *Corros. Sci.* **2010**, *52*, 2571–2581. [CrossRef]
109. Liu, W.; He, J.; Huang, H.; Wang, H.; Lu, Z.; Liu, C. Effects of Nb additions on the microstructure and mechanical property of CoCrFeNi high-entropy alloys. *Intermetallics* **2015**, *60*, 1–8. [CrossRef]
110. Shun, T.-T.; Hung, C.-H.; Lee, C.-F. Formation of ordered/disordered nanoparticles in FCC high entropy alloys. *J. Alloys Compd.* **2010**, *493*, 105–109. [CrossRef]
111. Shun, T.T.; Chang, L.Y.; Shiu, M.-H. Microstructures and mechanical properties of multiprincipal component $CoCrFeNiTi_x$ alloys. *Mater. Sci. Eng. A* **2012**, *556*, 170–174. [CrossRef]
112. Jiang, L.; Lu, Y.; Dong, Y.; Wang, T.; Cao, Z.; Li, T. Annealing effects on the microstructure and properties of bulk high-entropy CoCrFeNiTi 0.5 alloy casting ingot. *Intermetallics* **2014**, *44*, 37–43. [CrossRef]
113. Zhang, K.; Fu, Z. Effects of annealing treatment on phase composition and microstructure of $CoCrFeNiTiAl_x$ high-entropy alloys. *Intermetallics* **2012**, *22*, 24–32. [CrossRef]
114. Otto, F.; Yang, Y.; Bei, H.; George, E.P. Relative effects of enthalpy and entropy on the phase stability of equiatomic high-entropy alloys. *Acta Mater.* **2013**, *61*, 2628–2638. [CrossRef]
115. Cai, Z.; Jin, G.; Cui, X.; Li, Y.; Fan, Y.; Song, J. Experimental and simulated data about microstructure and phase composition of a NiCrCoTiV high-entropy alloy prepared by vacuum hot-pressing sintering. *Vacuum* **2016**, *124*, 5–10. [CrossRef]
116. Liu, L.; Zhu, J.; Zhang, C.; Li, J.; Jiang, Q. Microstructure and the properties of $FeCoCuNiSn_x$ high entropy alloys. *Mater. Sci. Eng. A* **2012**, *548*, 64–68. [CrossRef]
117. Samal, S.; Rahul, M.; Kottada, R.S.; Phanikumar, G. Hot deformation behaviour and processing map of Co-Cu-Fe-Ni-Ti eutectic high entropy alloy. *Mater. Sci. Eng. A* **2016**, *664*, 227–235. [CrossRef]
118. Kao, Y.-F.; Chen, S.-K.; Sheu, J.-H.; Lin, J.-T.; Lin, W.-E.; Yeh, J.-W.; Lin, S.-J.; Liou, T.-H.; Wang, C.-W. Hydrogen storage properties of multi-principal-component $CoFeMnTi_xVyZrz$ alloys. *Int. J. Hydrogen Energy* **2010**, *35*, 9046–9059. [CrossRef]
119. Jiang, L.; Cao, Z.; Jie, J.; Zhang, J.; Lu, Y.; Wang, T.; Li, T. Effect of Mo and Ni elements on microstructure evolution and mechanical properties of the $CoFeNi_xVMo_y$ high entropy alloys. *J. Alloys Compd.* **2015**, *649*, 585–590. [CrossRef]

120. Jiang, L.; Lu, Y.; Wu, W.; Cao, Z.; Li, T. Microstructure and Mechanical Properties of a CoFeNi2V0.5Nb0.75 Eutectic High Entropy Alloy in As-cast and Heat-treated Conditions. *J. Mater. Sci. Technol.* **2016**, *32*, 245–250. [CrossRef]
121. Zuo, T.; Li, R.; Ren, X.; Zhang, Y. Effects of Al and Si addition on the structure and properties of CoFeNi equal atomic ratio alloy. *J. Magn. Magn. Mater.* **2014**, *371*, 60–68. [CrossRef]
122. Han, Z.; Liu, X.; Zhao, S.; Shao, Y.; Li, J.; Yao, K. Microstructure, phase stability and mechanical properties of Nb–Ni–Ti–Co–Zr and Nb–Ni–Ti–Co–Zr–Hf high entropy alloys. *Prog. Nat. Sci. Mater. Int.* **2015**, *25*, 365–369. [CrossRef]
123. Jiang, H.; Zhang, H.; Huang, T.; Lu, Y.; Wang, T.; Li, T. Microstructures and mechanical properties of Co2MoxNi2VWx eutectic high entropy alloys. *Mater. Des.* **2016**, *109*, 539–546. [CrossRef]
124. Lee, C.; Chang, C.; Chen, Y.; Yeh, J.; Shih, H. Effect of the aluminium content of AlxCrFe1.5MnNi0.5 high-entropy alloys on the corrosion behaviour in aqueous environments. *Corros. Sci.* **2008**, *50*, 2053–2060. [CrossRef]
125. Kunce, I.; Polanski, M.; Bystrzycki, J. Structure and hydrogen storage properties of a high entropy ZrTiVCrFeNi alloy synthesized using Laser Engineered Net Shaping (LENS). *Int. J. Hydrogen Energy* **2013**, *38*, 12180–12189. [CrossRef]
126. Jiang, H.; Jiang, L.; Han, K.; Lu, Y.; Wang, T.; Cao, Z.; Li, T. Effects of tungsten on microstructure and mechanical properties of $CrFeNiV_{0.5}W_x$ and $CrFeNi_2V_{0.5}W_x$ high-entropy alloys. *J. Mater. Eng. Perform.* **2015**, *24*, 4594–4600. [CrossRef]
127. Fazakas, E.; Zadorozhnyy, V.; Varga, L.; Inoue, A.; Louzguine-Luzgin, D.; Tian, F.; Vitos, L. Experimental and theoretical study of $Ti_{20}Zr_{20}Hf_{20}Nb_{20}X_{20}$ (X= V or Cr) refractory high-entropy alloys. *Int. J. Refract. Met. Hard Mater.* **2014**, *47*, 131–138. [CrossRef]
128. Senkov, O.; Woodward, C. Microstructure and properties of a refractory NbCrMo0.5Ta0.5TiZr alloy. *Mater. Sci. Eng. A* **2011**, *529*, 311–320. [CrossRef]
129. Senkov, O.; Senkova, S.; Woodward, C.; Miracle, D. Low-density, refractory multi-principal element alloys of the Cr–Nb–Ti–V–Zr system: Microstructure and phase analysis. *Acta Mater.* **2013**, *61*, 1545–1557. [CrossRef]
130. Takeuchi, A.; Wada, T.; Zhang, Y. MnFeNiCuPt and MnFeNiCuCo high-entropy alloys designed based on L1 0 structure in Pettifor map for binary compounds. *Intermetallics* **2017**, *82*, 107–115. [CrossRef]
131. Guo, N.; Wang, L.; Luo, L.; Li, X.; Chen, R.; Su, Y.; Guo, J.; Fu, H. Microstructure and mechanical properties of refractory high entropy (Mo0.5NbHf0.5ZrTi)BCC/M5Si3 in-situ compound. *J. Alloys Compd.* **2016**, *660*, 197–203. [CrossRef]
132. Zhang, Y.; Liu, Y.; Li, Y.; Chen, X.; Zhang, H. Microstructure and mechanical properties of a refractory HfNbTiVSi0.5 high-entropy alloy composite. *Mater. Lett.* **2016**, *174*, 82–85. [CrossRef]
133. Wu, Y.; Cai, Y.; Chen, X.; Wang, T.; Si, J.; Wang, L.; Wang, Y.; Hui, X. Phase composition and solid solution strengthening effect in TiZrNbMoV high-entropy alloys. *Mater. Des.* **2015**, *83*, 651–660. [CrossRef]
134. Anzorena, M.S.; Bertolo, A.; Gagetti, L.; Kreiner, A.; Mosca, H.; Bozzolo, G.; del Grosso, M. Characterization and modeling of a MoTaVWZr high entropy alloy. *Mater. Des.* **2016**, *111*, 382–388. [CrossRef]

metals

MDPI

Article

Microstructure Evolution and Mechanical Properties of Refractory Mo-Nb-V-W-Ti High-Entropy Alloys

Maximilian Regenberg *, Georg Hasemann *, Markus Wilke, Thorsten Halle and Manja Krüger

Institute of Materials and Joining Technology, Otto-von-Guericke University Magdeburg, Universitätsplatz 2, 39106 Magdeburg, Germany; markus.wilke@ovgu.de (M.W.); thorsten.halle@ovgu.de (T.H.); manja.krueger@ovgu.de (M.K.)

* Correspondence: maximilian.regenberg@ovgu.de (M.R.); georg1.hasemann@ovgu.de (G.H.); Tel.: +49-391-67-54572 (M.R.)

Received: 23 September 2020; Accepted: 13 November 2020; Published: 18 November 2020

Abstract: High-entropy alloys can either be defined as solid solution alloys containing at least five elements in equiatomic or near-equiatomic composition, or as alloys with high configurational entropies (larger than 1.5R), regardless of the number of elements involved. The present study reports on an alloy design route for refractory high-entropy alloys based on equiatomic Mo-Nb-V alloys with additions of W and Ti. In general, the work was motivated by Senkov et al. The aim of the experiments carried out was to produce a refractory high-entropy alloy with a single-phase structure. For this purpose, a systematic alloy design involving four- and five-element compositions was used. Scanning electron microscopy analysis has shown that Mo-Nb-V-xW-yTi ($x = 0, 20$; $y = 5, 10, 15, 20, 25$) is in fact a refractory high-entropy alloy with a body-centered cubic dendritic structure. Furthermore, the Ti-concentration of the experimental alloys was varied, to obtain the influence of Titanium on the microstructure development. Additionally, compressive tests at room temperature were carried out to evaluate the influence of the different alloying elements and the Ti-fraction on the mechanical properties. The observations of the present work are then compared to the published results on similar alloys from the working group of Yao et al. and critically discussed.

Keywords: high-entropy alloys; refractory metals; microstructure; mechanical properties

1. Introduction

To meet the increasing demand for modern high-temperature materials, extensive research is being carried out in this field. Especially characteristics such as high-temperature strength, creep resistance and oxidation behavior are in the focus of present investigations. As the melting temperatures of materials correlate with the aforesaid properties, emphasis has been put on for example Fe, Ni or Pt and refractory elements like Mo and Nb, as basic alloying components in the manner of producing rather simple binary and ternary alloys, up to complex multi-component superalloys [1–4].

Besides the classical alloying strategies, starting from a base element and adding minor element additions to design and produce alloys, a multi-component alloy strategy offers the opportunity to design and investigate a completely new class of materials. In general, those alloys consist of five (or more) elements. They are referred to as high-entropy alloys (HEAs) if a single-phase material with relatively simple fcc (face-centered cubic), bcc (body centered cubic), or hcp (hexagonal close-packed) crystal structures is reached [5–8]. These, mostly equiatomic alloys, gained tremendous scientific attention, caused by their unique properties, such as high strength and hardness (even at high temperatures), outstanding wear resistance, good structural stability, corrosion and oxidation resistance. Those properties are attributed to certain interatomic and lattice effects: the core effect, high entropy effect, sluggish diffusion effect, severe-lattice-distortion effect and a cocktail effect, which are described in detail elsewhere [5,8]. HEAs can be seen as a scientific approach to investigate and understand the metal–physical nature of this new type of materials.

A more practical (or engineering) approach on HEAs may lead to near-equiatomic multi-component alloys in which the element composition can vary in a certain compositional interval (between 5 at.% and 35 at.% as it is defined by Yeh [9]), for each principle element. As a result of that, it is possible to design so-called compositionally complex alloys (CCAs) which show the potential to lead to new developments regarding types of superalloys in which a multi-component solid solution matrix contains high amounts of complex precipitation phases [10–13].

During the last decade HEAs based on Fe, Al, Cr, Ni, Co, Cu and Mn were emphasized and extensively examined. A well-studied example of this development is Fe-Cr-Mn-Ni-Co [14–17], which is also known as the so-called Cantor alloy. To be noted in addition is the US Patent published by Bei [18] in which a multi-component solid solution alloy with high mixing entropy, namely Mo-Nb-V-W-Ti, has been mentioned. Senkov et al. [19,20] introduced HEAs based primarily on refractory metals (RHEAs) in demand for high- and ultra-high-temperature structural materials, i.e., in the aerospace industry. Their alloys based on W-Nb-Mo-Ta and W-Nb-Mo-Ta-V were experimentally investigated and have a dendritic bcc structure, which, compared to the Ni-based alloy Inconel 718, shows relatively high yield stresses in a wide temperature range between 600 °C and 1600 °C [20]. These observations lead to the assumption, that the patented alloy Mo-Nb-V-W-Ti also consists of a single-phase body-centered cubic structure. Senkov et al.'s alloys, however, have relatively high densities of 12.36 g/cm^3 for W-Nb-Mo-Ta and 13.75 g/cm^3 for W-Nb-Mo-Ta-V [19].

In general, the present work aims to follow Senkov et al.'s approach of developing an RHEA, based on principle refractory metals but by substituting at least one and/or both of the heavy elements W and Ta with Ti, to reduce the density of the RHEAs significantly. Thus, the alloying strategy of the present work is based on theoretical calculations which give rise to a stable solid solution formation. In addition, the effects of removing/substituting aforesaid elements on the strength and ductility of the alloys at room temperature was examined.

2. Materials and Methods

For the alloy design of the present study, three quaternary alloys $Mo_{25}Nb_{25}V_{25}W_{25}$, $Mo_{25}Nb_{25}V_{25}Ti_{25}$, $Mo_{30}Nb_{30}V_{30}Ti_{10}$ and two quinary alloys $Mo_{20}Nb_{20}V_{20}W_{20}Ti_{20}$, $Mo_{22.5}Nb_{22.5}V_{22.5}W_{22.5}Ti_{10}$ were chosen (all concentrations are given in atomic percent). The alloys were prepared from high-purity metal chips, granules and powders via arc-melting under protective (Ar) atmosphere (Arc Melter MAM-1, Edmund Buehler GmbH, Bodelshausen, Germany). Button-shaped samples of 5–10 g were re-melted and flipped five times each, to ensure good homogeneity.

Powder samples of the alloys produced were manually prepared for X-ray diffraction analysis (XRD, X'Pert PRO, Malvern Panalytival B.V., Almelo, The Netherlands), using a hand file. The measuring parameters chosen for the XRD analysis were Cu-Kα_1-radiation (λ = 1.541874 Å) at a voltage of 40 kV, current intensity of 30 mA and increment size of 0.0167°.

In order to characterize the as-cast microstructure, the button-shaped samples of each alloy were cut and prepared for metallographic examination. After initial grinding, the specimens were polished using a 3 μm and a 1 μm diamond suspension, and finally polished with colloidal silica (OP-S, Struers, Ballerup, Denmark). The microstructural observations were carried out using a scanning electron microscope (SEM) (ESEM XL30 FEG, FEI, Hillsboro, OR, USA). Additionally, to validate the chemical composition of the alloy samples, (Si(Li))-detector Energy-dispersive X-ray spectroscopy (EDS) analysis, equipped with Genesis software (EDAX, Mahwah, NJ, USA) was conducted. Moreover, area fractions of the microstructures of the Ti-containing 5-element alloys $Mo_{20}Nb_{20}V_{20}W_{20}Ti_{20}$ and $Mo_{22.5}Nb_{22.5}V_{22.5}W_{22.5}Ti_{10}$ were calculated by means of graphical image analysis (Photoshop, Adobe, San José, CA, USA) using 15 images each.

Furthermore, Vickers hardness measurements with an applied force of 100 N were performed on the prepared specimens, using a hardness testing device (Duravision, Struers, Ballerup, Denmark).

Regarding the compression tests, 3 to 4 cylindrical samples conforming to standards [21] with specifications of 3 mm in length and 1.8 mm in diameter were prepared by electric discharge machining (EDM). The samples were surface-cleaned (removal of a burr and the surface tarnish) subsequently afterwards, using 800 and 1200 grit SiC paper. The compression tests were performed on an electro-mechanical universal testing machine (TIRAtest 2825, TIRA GmbH, Schalkau, Germany) at a constant crosshead speed corresponding to an initial (engineering) strain rate of $\dot{\varepsilon} = 10^{-3}\ s^{-1}$, up to a maximum deformation of ~50%. The yield strength was determined by the 0.2% offset method and the plastic strain was obtained by subtracting the combined compliance of the testing machine and the sample from the individually measured load-displacement curves.

3. Results and Discussion

3.1. Alloy Design Based on Thermodynamics and Geometrical Effects

The formation of a solid solution phase in multi-component alloys is determined by the system's Gibbs free energy ΔG_{mix} [22],

$$\Delta G_{mix} = \Delta H_{mix} - T\Delta S_{mix} \quad (1)$$

Hence, the phase formation in RHEAs is strongly dependent upon the interplay between the entropy of mixing ΔS_{mix}, the absolute temperature T and the enthalpy of mixing ΔH_{mix} [5]. Therefore, the following equations were used to determine the phase stability of the alloys examined [15,16]:

$$\Delta S_{mix} = -R\sum_{i=1}^{N} x_i \ln x_j \quad (2)$$

$$\Delta H_{mix} = \sum_{i=1;\, i\neq j}^{N} 4\Delta H_{AB}^{mix} x_i x_j \quad (3)$$

where, R is the gas constant, 8.314 $J{\cdot}mol^{-1}{\cdot}K^{-1}$, N is the number of elements, x_i or x_j is the composition of the ith or jth element and ΔH_{AB}^{mix} is the enthalpy of mixing for the binary equiatomic AB alloys. Furthermore, the ratio of entropy to enthalpy Ω (also referred to as solid solution prediction parameter) was used to make predictions about solid solution formation in the present alloying systems [23]:

$$\Omega = \frac{\sum_{i=1}^{n} c_i (T_m)_i \Delta S_{mix}}{|\Delta H_{mix}|} \quad (4)$$

here, the melting temperature of an n-element alloy is calculated, where T_m is used as the melting point of the ith component of the alloy. Additionally, the atomic size difference δ_r in % was taken into account, using the equation given below [24]:

$$\delta_r\% = 100\% \sqrt{\sum_{i=1}^{n} c_i \left(1 - \frac{r_i}{\bar{r}}\right)^2} \quad (5)$$

where, C_i is the atomic fraction of the ith component, $\bar{r} = \sum_{i=1}^{n} c_i r_i$ is the average atomic radius and r_i is the atomic radius (listed in reference [25]).

The results of the calculations performed (as displayed in Table 1) are in good agreement with the findings of Zhang et al. [26,27], stating that low values of ΔH_{mix} and, respectively, high values of ΔS_{mix} reduce the free energy of the system, thus providing a stable solid solution phase. This assumption is supported by the Ω values stated, all of which are significantly higher than 1 ($\Omega_{min} = \Omega(Mo_{25}Nb_{25}V_{25}W_{25}) = 8.27$), leading to a more dominant contribution of $(T)\Delta S_{mix}$ than of ΔH_{mix} to the free energy of the system [23]. In addition to the entropy of mixing, the δ_r data of the elements mixed contribute to the formation of phases, since the lattice distortion is influenced by the different atomic radii [28]. The stated RHEA alloys show δ_r values between 2.98% and 4.62%, hence indicating the formation of a stable solid solution [29].

Table 1. Investigated alloys, and calculated values for the entropy of mixing ΔS_{mix}, the enthalpy of mixing ΔH_{mix}, the solid solution prediction parameter Ω and the atomic size difference δ_r.

Alloy	ΔS_{mix}, $J \cdot K^{-1} \cdot mol^{-1}$	ΔH_{mix}, $kJ \cdot mol^{-1}$	Ω	δ_r, %
$Mo_{25}Nb_{25}V_{25}W_{25}$	11.52	−4.00	8.27	2.98
$Mo_{25}Nb_{25}V_{25}Ti_{25}$	11.52	−2.75	10.20	4.46
$Mo_{30}Nb_{30}V_{30}Ti_{10}$	10.92	−3.00	9.23	4.01
$Mo_{20}Nb_{20}V_{20}W_{20}Ti_{20}$	13.38	−4.16	8.63	4.05
$Mo_{22.5}Nb_{22.5}V_{22.5}W_{22.5}Ti_{10}$	13.08	−4.14	8.77	3.65

To demonstrate and evaluate the potential of the alloys examined, the generated ΔH_{mix} vs. δ_r data were graphically displayed in Figure 1a by means of a field diagram, showing the regions of theoretical solid solution, respectively, amorphous phase formation, and compared to obtained data from the corresponding literature.

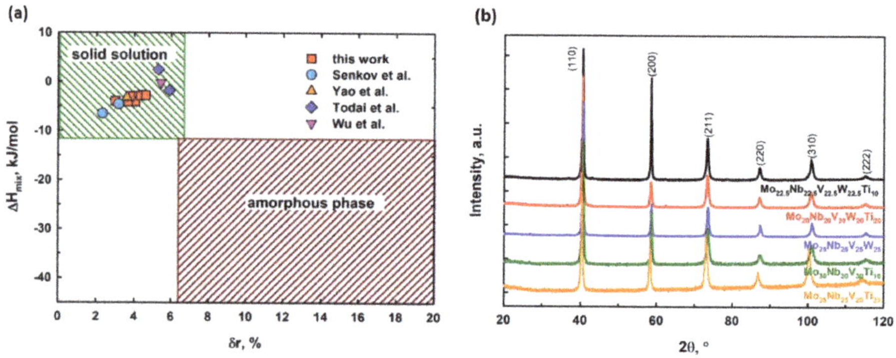

Figure 1. (**a**) Calculated ΔH_{mix} to δ_r ratios from the present work compared to data, obtained from the corresponding literature [26,27]. (**b**) XRD patterns of the alloys examined in the as-cast condition (from top to bottom).

3.2. X-ray Diffraction of the RHEAs

The XRD analysis results of the RHEA powder samples are depicted in patterns, displaying the relative intensity of the detected reflexes over a 2-θ-range from 20° to 120° in Figure 1b. The XRD reflexes of the different alloys show similar locations, indicating the presence of a bcc crystal structure in all alloys examined. The positions of the diffraction reflexes present are characteristic for elements with a bcc crystal structure (f.e. the refractory elements Mo, V, Nb, W, etc.) in general, however they are slightly shifted here [30]. The exact position of the reflexes obtained varies, due to the lattice distortion of the different solid solutions. The experimentally obtained lattice parameters reaching from a minimum of 3.156 Å ($Mo_{22.5}Nb_{22.5}V_{22.5}W_{22.5}Ti_{10}$) to a maximum value of 3.174 Å ($Mo_{25}Nb_{25}V_{25}Ti_{25}$) are listed in Table 2. Theoretical values for the alloys examined were determined, using the equation [19]:

$$a_{theor.} = \sum c_i a_i \tag{6}$$

and the results, stated in Table 2 are in good agreement with the experimental values.

Table 2. Theoretically calculated and experimentally measured lattice parameters of the alloys investigated with standard deviations.

Alloy	Theor. Lattice Parameter, Å	Exp. Lattice Parameter, Å	Deviation
$Mo_{25}Nb_{25}V_{25}W_{25}$	3.160	3.157	±0.11
$Mo_{25}Nb_{25}V_{25}Ti_{25}$	3.196	3.174	±0.009
$Mo_{30}Nb_{30}V_{30}Ti_{10}$	3.173	3.156	±0.002
$Mo_{20}Nb_{20}V_{20}W_{20}Ti_{20}$	3.189	3.164	±0.013
$Mo_{22.5}Nb_{22.5}V_{22.5}W_{22.5}Ti_{10}$	3.175	3.159	±0.008

3.3. *As-Cast Microstructures of the Alloys Examined*

As-cast microstructure overview-images of the five alloys produced are depicted in Figure 2. The images were taken at a central position of the arc-melted buttons. In addition, the results of the chemical compositions, analyzed by means of EDS, are shown in Table 3. It can be stated that the actual element concentration is in good agreement with the theoretical calculated values, with a maximum deviation of ±2.1 at.% W in alloy $Mo_{20}Nb_{20}V_{20}W_{20}Ti_{20}$, regarding all alloys tested. The microstructure evolution is of a similar kind within the button-shaped samples. In principal, the formation of dendrites in different sizes and refinement can be examined. As displayed on alloy $Mo_{25}Nb_{25}V_{25}W_{25}$ in Figure 3, a coarse-grained dendritic structure is present in the bottom regions of the button-shaped samples, while a more distinctive growth can be obtained from the mid to the top section of the samples. This is attributed to the cooling conditions in the arc-melter or more specifically the fast heat transfer from the sample into the water-cooled copper crucible, thus restricting the dendrite growth.

Even though theory predicts the formation of a solid solution, regarding the Ti-containing 5-element alloys, the presence of a second phase was observed at higher magnifications (small, black colored areas), which are depicted in the micrograph insets of the EDS element mappings. The EDS analysis (Figure 4) revealed that these areas are indeed Ti-rich precipitations, which form at the grain boundaries of the dendrites during solidification (~5% area fraction in alloy $Mo_{22.5}Nb_{22.5}V_{22.5}W_{22.5}Ti_{10}$ and ~7% area fraction in alloy $Mo_{20}Nb_{20}V_{20}W_{20}Ti_{20}$). The reason for the formation of the dendritic microstructure in general and the Ti-precipitations in particular lies within the different melting temperatures of the elements present. As titanium exhibits the lowest melting temperature ($T_m \approx 1660$ °C) of the alloying elements used, hence solidifying last, the solidification front is pushing the low viscous residual melt forward, resulting in a Ti-enriched interdendritic zone [31]. This observation can also be expanded to vanadium, which exhibits the second lowest melting temperature ($T_m \approx 1910$ °C), thus experiencing the same influence (depicted in the EDS mappings in Figure 4). Depending on the alloy composition, tungsten ($T_m \approx 3420$ °C), molybdenum ($T_m \approx 2620$ °C) and niobium ($T_m \approx 2480$ °C) constitute the dendritic crystals, respectively, suppressing the remaining alloying elements in the interdendritic region. If W is present in the alloy (cf. the five-component alloy (Figure 4b), the primary dendrites mainly consist of it, which is due to the extremely high melting point. If the alloy only contains Mo and Nb instead of W as high melting components (cf. the four-element alloy (Figure 4a), the primary dendrites are enriched with both of the aforesaid elements, which can be attributed to the more similar meting point of the components. The results of the EDS mappings were quantitatively verified by means of spot analysis in the dendritic (C_d), interdendritic (C_{id}) and precipitation (C_{pr}) regions of the samples (examples regarding alloy $Mo_{25}Nb_{25}V_{25}Ti_{25}$ and $Mo_{20}Nb_{20}V_{20}W_{20}Ti_{20}$ are listed in Table 4) and are overall in good agreement.

Table 3. Chemical compositions (at.%) of the alloys investigated with standard deviations, determined by EDS analysis.

Alloy	Mo	V	Nb	W	Ti
$Mo_{25}Nb_{25}V_{25}W_{25}$	28 ± 1	21 ± 1	26 ± 1	25 ± 2	-
$Mo_{25}Nb_{25}V_{25}Ti_{25}$	26 ± 1	24 ± 0.1	26 ± 0.1	-	24 ± 1
$Mo_{30}Nb_{30}V_{30}Ti_{10}$	31 ± 0.1	28 ± 0.1	31 ± 0.1	-	10 ± 0.2
$Mo_{20}Nb_{20}V_{20}W_{20}Ti_{20}$	21 ± 1	19 ± 1	21 ± 0.4	20 ± 2	19 ± 1
$Mo_{22.5}Nb_{22.5}V_{22.5}W_{22.5}Ti_{10}$	26 ± 1	23 ± 0.2	22 ± 0.2	20 ± 0.2	10 ± 0.1

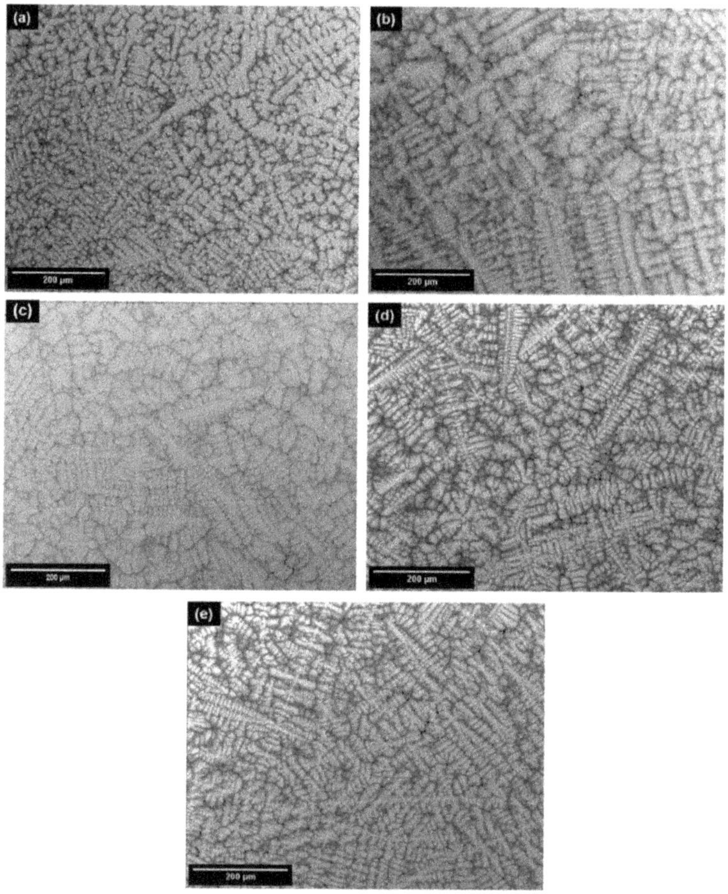

Figure 2. SEM backscattered electron (BSE) overview-images of the as-cast microstructures of $Mo_{25}Nb_{25}V_{25}W_{25}$ (**a**), $Mo_{25}Nb_{25}V_{25}Ti_{25}$ (**b**), $Mo_{30}Nb_{30}V_{30}Ti_{10}$ (**c**), $Mo_{20}Nb_{20}V_{20}W_{20}Ti_{20}$ (**d**) and $Mo_{22.5}Nb_{22.5}V_{22.5}W_{22.5}Ti_{10}$ (**e**).

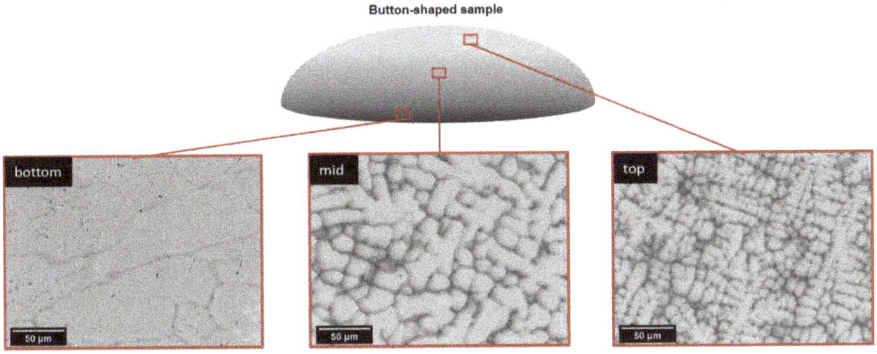

Figure 3. Microstructural differences in the formation of dendrites across the produced button-shaped samples, exemplified on alloy $Mo_{25}Nb_{25}V_{25}W_{25}$.

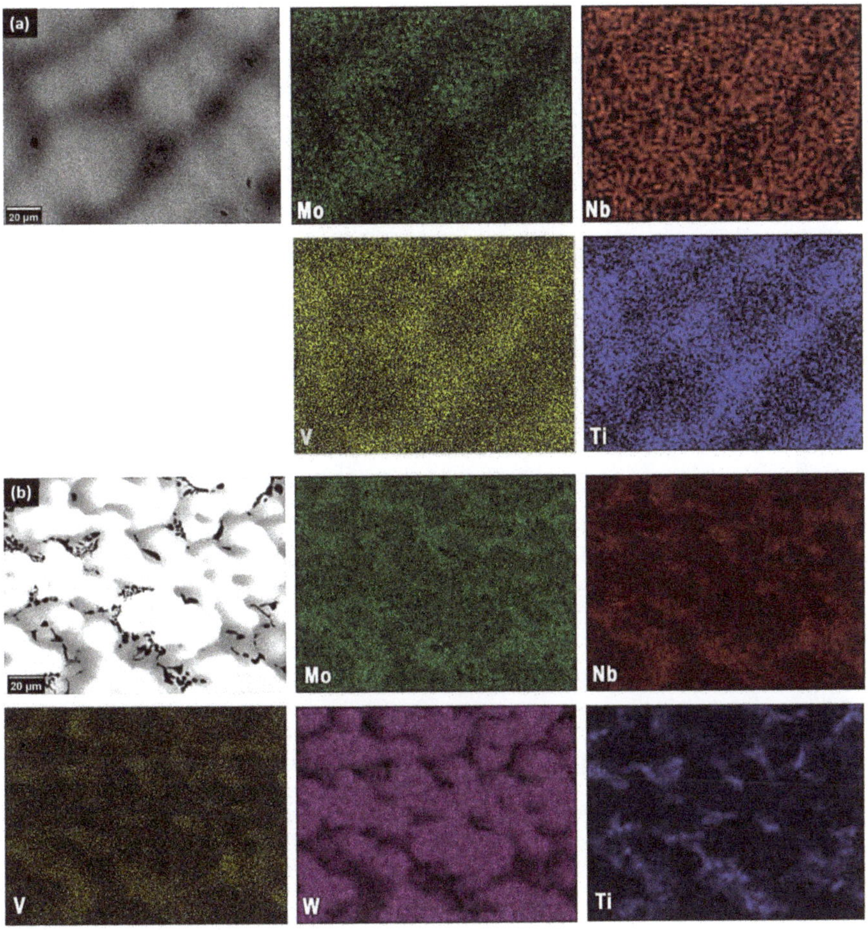

Figure 4. EDS mappings of the refractory high-entropy alloys (RHEAs) $Mo_{25}Nb_{25}V_{25}Ti_{25}$ (**a**) and $Mo_{20}Nb_{20}V_{20}W_{20}Ti_{20}$ (**b**).

Table 4. Results of the EDS spot analysis in the dendritic (C_d), interdendritic (C_{id}), and precipitation (C_{pr}) areas of alloy $Mo_{25}Nb_{25}V_{25}Ti_{25}$ and alloy $Mo_{20}Nb_{20}V_{20}W_{20}Ti_{20}$.

Alloy	Area	Mo, at.%	Nb, at.%	V, at.%	W, at.%	Ti, at.%
$Mo_{25}Nb_{25}V_{25}Ti_{25}$	C_d	28	28	22	-	22
	C_{id}	16	22	31	-	30
$Mo_{20}Nb_{20}V_{20}W_{20}Ti_{20}$	C_d	23	19	14	32	12
	C_{id}	19	25	26	11	19
	C_{pr}	2	7	5	2	85

3.4. Mechanical Properties at Room Temperature

The results of the compression tests regarding the obtained compressive yield strength values (yield strength in the following) vs. the theoretical density of the samples, determined by the formula to calculate the density of a disordered solid solution [19]:

$$\rho_{\text{theor.}} = \frac{\sum c_i A_i}{\sum \frac{c_i A_i}{\rho_i}}, \tag{7}$$

are displayed in Figure 5a. In addition, the obtained hardness values of the RHEAs vs. the theoretical density are depicted in Figure 5b. The corresponding numerical values as well as the calculated theoretical densities of the alloys examined are stated in Table 5. Overall, the alloys tested show relatively high yield strength and low ductility values, which is attributed to the large valence electron concentration (VEC) [8,32] and are in good agreement with the findings of Senkov et al. [19,20]. Nonetheless, it must be stated that the values considering the compressive yield stress partly show some major deviations which is attributed to the presence of porosity in the compression samples. This originates mainly in the manufacturing process. As the specimens were produced by means of arc melting, process related porosity could not be prevented. The ultra-high melting points of the elements used complicate the melting process, leading to a highly viscous melt which solidifies quickly, thus resulting in the present gas porosity and cavities. Referring to the samples of alloy $Mo_{22.5}Nb_{22.5}V_{22.5}W_{22.5}Ti_{10}$, the obtained porosity is so significant that it can be determined as the main failure mechanism, thus the results of alloy $Mo_{22.5}Nb_{22.5}V_{22.5}W_{22.5}Ti_{10}$ were not taken into account and are therefore not displayed in the images. By comparing the depicted graphs, it is noticeable that there is no linear dependency present between the yield strength and the theoretical density of the alloys in contrast to the measured hardness values and the theoretical density. The yield strength values of the alloys $Mo_{25}Nb_{25}V_{25}Ti_{25}$, $Mo_{20}Nb_{20}V_{20}W_{20}Ti_{20}$ and $Mo_{25}Nb_{25}V_{25}W_{25}$ are comparable, while alloy $Mo_{30}Nb_{30}V_{30}Ti_{10}$ displays the highest value. This observation leads to the assumption that a reduced Ti-content (10 at.% in alloy $Mo_{30}Nb_{30}V_{30}Ti_{10}$ vs. 25 at.% in alloy $Mo_{25}Nb_{25}V_{25}Ti_{25}$ and 20 at.% in alloy $Mo_{20}Nb_{20}V_{20}W_{20}Ti_{20}$), respectively, and higher content of the elements Mo, Nb and V, results in a higher compressive yield strength. By showing a relatively low yield strength, however, alloy $Mo_{25}Nb_{25}V_{25}W_{25}$ is in contrast to this statement. On the one hand, this can be attributed to the aforementioned porosity in the $Mo_{25}Nb_{25}V_{25}W_{25}$ specimens. On the other hand, the high W (25 at.% W) concentration in the alloy leads to lower concentrations of the strength enhancing components Mo, Nb and V (compared to alloy $Mo_{30}Nb_{30}V_{30}Ti_{10}$ f.e.) and the enrichment of W in the dendritic crystals must be considered. This results in a suppression of the lower-melting elements in the interdendritic regions, hence generating an imbalance in the element distribution.

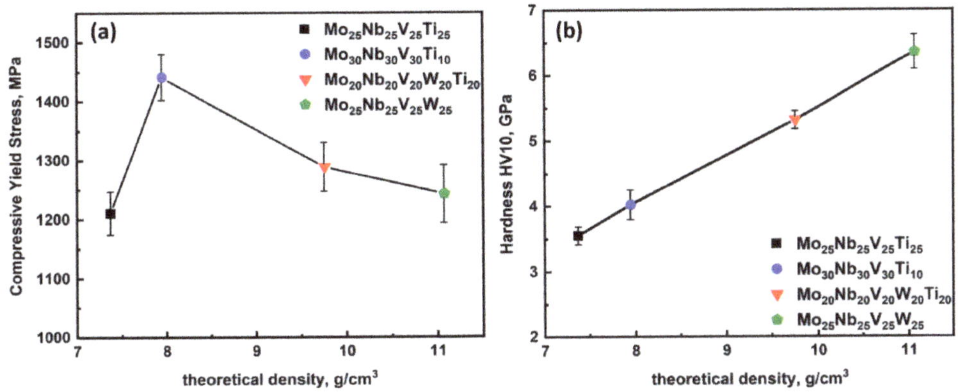

Figure 5. Compressive Yield Stress—theoretical density curve (**a**) and hardness—theoretical density curve (**b**) of the alloys investigated.

Table 5. Theoretical densities of the present HEAs in increasing order, as well as the obtained compressive yield stress values, the specific compressive yield stress values and the hardness values.

Alloy	$\rho_{theor.}$, g·cm^{-3}	$\sigma_{0.2}$, MPa	$\sigma_{0.2\ spec.}$, MPa·cm^3/g	Hardness HV10, GPa
$Mo_{25}Nb_{25}V_{25}Ti_{25}$	7.4	1210 ± 36	164 ± 5	4 ± 0.1
$Mo_{30}Nb_{30}V_{30}Ti_{10}$	7.9	1441 ± 39	182 ± 5	4 ± 0.2
$Mo_{20}Nb_{20}V_{20}W_{20}Ti_{20}$	9.8	1289 ± 42	132 ± 4	5 ± 0.1
$Mo_{25}Nb_{25}V_{25}W_{25}$	11.1	1243 ± 49	112 ± 4	6 ± 0.2

Regarding the macro-hardness of the samples and the density, a clear connection between the alloying elements can be obtained. Alloy $Mo_{25}Nb_{25}V_{25}Ti_{25}$ displays the lowest HV10 value as well as the lowest density, due to the high fraction of Ti. With decreasing Ti content in relation to the other alloying elements, the hardness of the microstructure increases in correlation with the density. A significant increase (both in hardness and density) can be examined regarding the alloys $Mo_{20}Nb_{20}V_{20}W_{20}Ti_{20}$ and $Mo_{25}Nb_{25}V_{25}W_{25}$, respectively, which is attributed to the addition of W [19]. Compared to Mo-Nb-V-Ta-Ti alloys investigated by Bei et al. [18] the present results are in good agreement with their hardness values. Thus, substituting W with Ta reveals similar mechanical properties in combination with a decreased density and melting point of the alloy.

In order to position the yield strength in relation to the density (hence the W fraction as it affects the alloys examined the most), the specific compressive yield strengths (SCYS) of the alloys tested were calculated (Table 5). The calculated values were then compared to the obtained compressive yield stresses and displayed in Figure 6. The graph clearly demonstrates the influence of the light elements (mainly Ti, but also V), respectively, the heavy element (W) in the alloy composition: Alloy $Mo_{25}Nb_{25}V_{25}Ti_{25}$ exhibits a major gap between the two values, which is attributed to the proportionally high Ti and V content, resulting in a comparatively high SCYS. Alloy $Mo_{30}Nb_{30}V_{30}Ti_{10}$ also shows a, however smaller, increase in the SCYS which still is the highest value overall (due to the aforementioned high fractions of Mo and Nb). In contrast to the alloys discussed, alloy $Mo_{20}Nb_{20}V_{20}W_{20}Ti_{20}$ and alloy $Mo_{25}Nb_{25}V_{25}W_{25}$ show a decrease in SCYS in relation to the obtained compressive yield stress, primarily caused by the increased density of both alloys, thus putting the actual increase in the compressive yield stress into perspective. Comparing the SCYS values of all alloys it can be stated that an approximation to a linear trend is present (in contrast to the compressive yield stress values), which leads to an interesting perspective and more faceted way of comparing the microstructure evolution and the resulting properties of RHEAs such as the ones investigated.

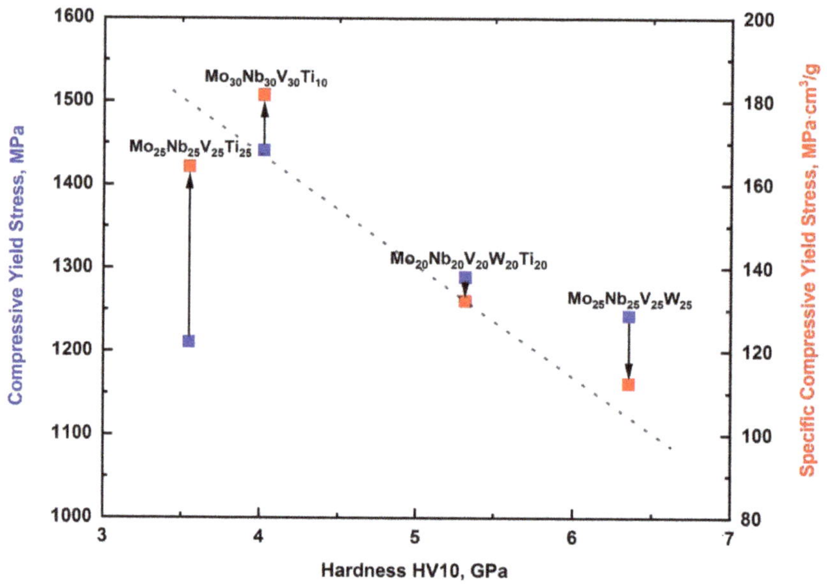

Figure 6. Comparison of compressive yield stress and specific compressive yield stress with respect to the hardness of the alloys tested.

3.5. Comparison with Other Works and Superalloys

A comparison regarding SCYS vs. the hardness of different RHEAs, which were the focus of published research in recent years by Yao et al. [33,34] and Senkov et al. [20], common Ni-based superalloys Inconel 718 [35], alloy CMSX-4 [36,37] and alloy Haynes 282 [38] and the alloys investigated in this work is displayed in Figure 7. Considering the SCYS and the hardness values, the examined alloys are superior in comparison to the superalloys. Only alloy $Mo_{25}Nb_{25}V_{25}W_{25}$ shows a lower SCYS, compared to alloy Inconel 718, however the hardness is significantly higher. In general, the examined alloys fit in well with the values obtained from the other RHEAs. The Ti-containing alloys $Mo_{25}Nb_{25}V_{25}Ti_{25}$ and $Mo_{30}Nb_{30}V_{30}Ti_{10}$ show the highest SCYS values of all alloys depicted, which can be attributed to the high Ti and V fractions, thus showing reduced density, compared with the Ta containing alloys from Yao et al. [33,34]. In comparison, the high W-containing alloy $Mo_{25}Nb_{25}V_{25}W_{25}$ has the lowest SCYS but highest hardness values of the alloys investigated in this study, similar to alloy $Mo_{20}Nb_{20}V_{20}W_{20}Ta_{20}$ by Senkov et al. [20]. Thus, the alloy $Mo_{20}Nb_{20}V_{20}W_{20}Ti_{20}$ represents a good compromise between strength and hardness which seems to agree with similar alloy compositions investigated by Yao et al. [33,34].

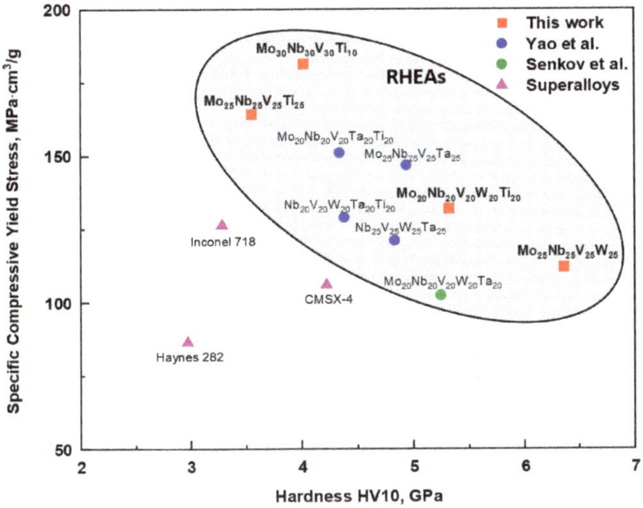

Figure 7. Specific compressive yield stress and hardness of different RHEAs, investigated by Yao et al. [33,34] and Senkov et al. [20], common Ni-based superalloys [35–38] and the alloys examined in this work for comparison.

4. Conclusions

The microstructure evolution and mechanical properties of RHEA alloys $Mo_{25}Nb_{25}V_{25}W_{25}$, $Mo_{25}Nb_{25}V_{25}Ti_{25}$, $Mo_{30}Nb_{30}V_{30}Ti_{10}$, $Mo_{20}Nb_{20}V_{20}W_{20}Ti_{20}$ and $Mo_{22.5}Nb_{22.5}V_{22.5}W_{22.5}Ti_{10}$ were investigated in the present work.

The alloy design was based on thermodynamic calculations regarding the Gibbs free energy ΔG_{mix}, the entropy of mixing ΔS_{mix}, the enthalpy of mixing ΔH_{mix} and the solid solution prediction parameter Ω, as well as geometrical effects, considering the difference of atomic radii δ_r.

XRD analysis showed that all alloys examined have a bcc crystal structure with a lattice parameter between 3.156 Å ($Mo_{22.5}Nb_{22.5}V_{22.5}W_{22.5}Ti_{10}$) and 3.174 Å ($Mo_{25}Nb_{25}V_{25}Ti_{25}$), however, the presence of a second, Ti-enriched phase could be detected subsequently via SEM observations. The second phase was not distinguished by the XRD method (no shoulders or modifications of the reflexes as such were determined) due to the small volume fractions present.

Microstructure analysis demonstrated that the alloys produced show a similar single-phase dendritic structure, apart from aforementioned alloy $Mo_{20}Nb_{20}V_{20}W_{20}Ti_{20}$ and alloy $Mo_{22.5}Nb_{22.5}V_{22.5}W_{22.5}Ti_{10}$. Deviations in the microstructure-refinement of the different samples were also obtained, which could be attributed to the cooling conditions during the arc-melting process.

To verify the composition of the samples, EDS spot analysis and mappings were conducted, which showed a good distribution of the alloying elements overall and an enrichment of high melting elements in the dendrite crystals vs. an accumulation of elements with a lower melting temperature in the interdendritic regions.

The mechanical properties of the alloys were obtained by means of room temperature compression tests and macro-hardness measurements, which revealed various impacts of the elements involved. The compressive yield strength of the alloys tested is mainly influenced by the Mo and Nb fraction present (see f.e. alloy $Mo_{30}Nb_{30}V_{30}Ti_{10}$: $\sigma_{0.2}$ = 1440 ± 40 MPa vs. alloy $Mo_{25}Nb_{25}V_{25}Ti_{25}$: $\sigma_{0.2}$ = 1210 ± 36 MPa). The W fraction primarily influences the hardness of the alloys; nonetheless, an impact on the compressive yield stress can be obtained regarding alloy $Mo_{20}Nb_{20}V_{20}W_{20}Ti_{20}$. The density of the Alloys examined was primarily determined by Ti, V and W (see alloy $Mo_{25}Nb_{25}V_{25}Ti_{25}$: highest share of Ti and V, lowest density vs. alloy $Mo_{25}Nb_{25}V_{25}W_{25}$: highest share of W, highest density). At this point, the determined porosity which was mainly caused by the arc-melting process as such and the highly viscous melt should also be noted. A different melting process (f.e. cold crucible induction melting) as well as the selection of lower-melting elements shall be mentioned here as alternatives for samples with reduced porosity/cavities.

Additionally, a comparison between the specific compressive yield strength and the nominal compressive yield strength has been carried out, which led to a more differentiated relation between the density (respectively, the content of light elements) and the compressive yield stress vs. the hardness of the alloys investigated. The calculated specific compressive yield stress values were finally compared to other RHEAs and Ni-base superalloys by means of a scatter diagram.

Author Contributions: Conceptualization, M.K., T.H. and G.H.; methodology, M.R., G.H. and M.W.; validation, G.H. and M.K.; formal analysis, M.R., G.H., M.W., T.H. and M.K.; investigation, M.R., G.H. and M.W.; writing—original draft preparation, M.R.; writing—review and editing; M.R., G.H. and M.K.; supervision, M.K. and T.H., funding acquisition, M.K. and T.H. All authors have read and agreed to the published version of the manuscript.

Funding: Financial support of the Methodisch-Diagnostisches Zentrum Werkstoffprüfung (MDZWP) e.V., Magdeburg, Germany and the "Graduiertenförderung Land Sachsen-Anhalt" is greatly acknowledged.

Acknowledgments: Experimental support by J. Kühn and M. Hellriegel is greatly acknowledged. M.R. would like to kindly thank Janett Schmelzer and Julia Becker for their support and fruitful discussions on the topic.

Conflicts of Interest: The authors declare no conflict of interest.

References

1. Speich, G.R.; Schwoeble, A.J.; Leslie, W.C. Elastic constants of binary iron-base alloys. *Metall. Trans.* **1972**, *3*, 2031–2037. [CrossRef]
2. Tawancy, H.M.; Klarstrom, D.L.; Rothman, M.F. Development of a New Nickel-Base Superalloy. *JOM* **1984**, *36*, 58–62. [CrossRef]
3. Hemmersmeier, U.; Feller-Kniepmeier, M. Element distribution in the macro- and microstructure of nickel base superalloy CMSX-4. *Mater. Sci. Eng. A* **1998**, *248*, 87–97. [CrossRef]
4. Dimiduk, D.M.; Perepezko, J.H. Mo-Si-B alloys: Developing a revolutionary turbine-engine material. *MRS Bull.* **2003**, *28*, 639–645. [CrossRef]
5. Gao, M.C.; Liaw, P.K.; Yeh, J.W.; Zhang, Y. *High-Entropy Alloys: Fundamentals and Applications*; Springer International Publishing: Basel, Switzerland, 2016; ISBN 9783319270135.
6. Yeh, J.W. Alloy design strategies and future trends in high-entropy alloys. *JOM* **2013**, *65*, 1759–1771. [CrossRef]
7. George, E.P.; Raabe, D.; Ritchie, R.O. High-entropy alloys. *Nat. Rev. Mater.* **2019**, *4*, 515–534. [CrossRef]
8. George, E.P.; Curtin, W.A.; Tasan, C.C. High entropy alloys: A focused review of mechanical properties and deformation mechanisms. *Acta Mater.* **2020**, *188*, 435–474. [CrossRef]
9. Tsai, M.H.; Yeh, J.W. High-entropy alloys: A critical review. *Mater. Res. Lett.* **2014**, *2*, 107–123. [CrossRef]
10. Manzoni, A.; Daoud, H.; Mondal, S.; Van Smaalen, S.; Völkl, R.; Glatzel, U.; Wanderka, N. Investigation of phases in Al23Co15Cr 23Cu8Fe15Ni16 and Al 8Co17Cr17Cu8Fe17Ni 33 high entropy alloys and comparison with equilibrium phases predicted by Thermo-Calc. *J. Alloys Compd.* **2013**, *552*, 430–436. [CrossRef]
11. Manzoni, A.M.; Daoud, H.M.; Voelkl, R.; Glatzel, U.; Wanderka, N. Influence of W, Mo and Ti trace elements on the phase separation in Al8Co17Cr17Cu8Fe17Ni33 based high entropy alloy. *Ultramicroscopy* **2015**, *159*, 265–271. [CrossRef]
12. Manzoni, A.M.; Singh, S.; Daoud, H.M.; Popp, R.; Völkl, R.; Glatzel, U.; Wanderka, N. On the path to optimizing the Al-Co-Cr-Cu-Fe-Ni-Ti high entropy alloy family for high temperature applications. *Entropy* **2016**, *18*, 104. [CrossRef]
13. Bolbut, V.; Wessel, E.; Krüger, M. Phase stability and temperature-dependent compressive strength of a low-density Fe32.3Al29.3Cu11.7Ni10.8Ti15.9 alloy. *Scr. Mater.* **2018**, *150*, 54–56. [CrossRef]
14. Cantor, B.; Chang, I.T.H.; Knight, P.; Vincent, A.J.B. Microstructural development in equiatomic multicomponent alloys. *Mater. Sci. Eng. A* **2004**, *375–377*, 213–218. [CrossRef]
15. Yeh, J.W.; Chen, S.K.; Lin, S.J.; Gan, J.Y.; Chin, T.S.; Shun, T.T.; Tsau, C.H.; Chang, S.Y. Nanostructured high-entropy alloys with multiple principal elements: Novel alloy design concepts and outcomes. *Adv. Eng. Mater.* **2004**, *6*, 299–303. [CrossRef]

16. Otto, F.; Yang, Y.; Bei, H.; George, E.P. Relative effects of enthalpy and entropy on the phase stability of equiatomic high-entropy alloys. *Acta Mater.* **2013**, *61*, 2628–2638. [CrossRef]
17. Otto, F.; Dlouhý, A.; Pradeep, K.G.; Kuběnová, M.; Raabe, D.; Eggeler, G.; George, E.P. Decomposition of the single-phase high-entropy alloy CrMnFeCoNi after prolonged anneals at intermediate temperatures. *Acta Mater.* **2016**, *112*, 40–52. [CrossRef]
18. Bei, H. Multi-Component Solid Solution Alloys Having High Mixing Entropy. US Patent US9150945B2, 6 October 2015.
19. Senkov, O.N.; Wilks, G.B.; Miracle, D.B.; Chuang, C.P.; Liaw, P.K. Refractory high-entropy alloys. *Intermetallics* **2010**, *18*, 1758–1765. [CrossRef]
20. Senkov, O.N.; Wilks, G.B.; Scott, J.M.; Miracle, D.B. Mechanical properties of Nb25Mo25Ta 25W25 and V20Nb20Mo 20Ta20W20 refractory high entropy alloys. *Intermetallics* **2011**, *19*, 698–706. [CrossRef]
21. Materialprüfung, DIN-Normenausschuss. Gleitlager, DIN-Normenausschuss Wälz- und DIN 50106 Prüfung Metallischer Werkstoffe—Druckversuch bei Raumtemperatur. 2016. Available online: https://www.din.de/de/mitwirken/normenausschuesse/nawgl (accessed on 16 November 2020).
22. Hasegawa, M. *Thermodynamic Basis for Phase Diagrams*; Elsevier Ltd.: London, UK, 2013; Volume 1, ISBN 9780080969862.
23. Yang, X.; Zhang, Y. Prediction of high-entropy stabilized solid-solution in multi-component alloys. *Mater. Chem. Phys.* **2012**, *132*, 233–238. [CrossRef]
24. Fang, S.; Xiao, X.; Xia, L.; Li, W.; Dong, Y. Relationship between the widths of supercooled liquid regions and bond parameters of Mg-based bulk metallic glasses. *J. Non-Cryst. Solids* **2003**, *321*, 120–125. [CrossRef]
25. Kittel, C. *Introduction to Solid State Physics*, 8th ed.; Wiley John + Sons: Berkeley, CA, USA, 2008; ISBN 047141526X.
26. Zhang, Y.; Zhou, Y.J.; Lin, J.P.; Chen, G.L.; Liaw, P.K. Solid-solution phase formation rules for multi-component alloys. *Adv. Eng. Mater.* **2008**, *10*, 534–538. [CrossRef]
27. Zhang, Y.; Lu, Z.P.; Ma, S.G.; Liaw, P.K.; Tang, Z.; Cheng, Y.Q.; Gao, M.C. Guidelines in predicting phase formation of high-entropy alloys. *MRS Commun.* **2014**, *4*, 57–62. [CrossRef]
28. Wu, C.S.; Tsai, P.H.; Kuo, C.M.; Tsai, C.W. Effect of atomic size difference on the microstructure and mechanical properties of high-entropy alloys. *Entropy* **2018**, *20*, 967. [CrossRef]
29. Ye, Y.F.; Wang, Q.; Lu, J.; Liu, C.T.; Yang, Y. High-entropy alloy: Challenges and prospects. *Mater. Today* **2016**, *19*, 349–362. [CrossRef]
30. Becker, J.; Betke, U.; Wessel, E.; Krüger, M. Alloying effects in Mo-5X (X = Zr, Ti,V)—Microstructural modifications and mechanical properties. *Mater. Today Commun.* **2018**, *15*, 314–321. [CrossRef]
31. Sahm, P.R.; Egry, I.; Volkmann, T. *Schmelze, Erstarrung, Grenzflächen*; Sahm, P.R., Egry, I., Volkmann, T., Eds.; Springer Berlin Heidelberg: Berlin/Heidelberg, Germany, 1999; ISBN 978-3-540-41566-4.
32. Sheikh, S.; Shafeie, S.; Hu, Q.; Ahlström, J.; Persson, C.; Veselý, J.; Zýka, J.; Klement, U.; Guo, S. Alloy design for intrinsically ductile refractory high-entropy alloys. *J. Appl. Phys.* **2016**, *120*, 164902. [CrossRef]
33. Yao, H.; Qiao, J.W.; Gao, M.C.; Hawk, J.A.; Ma, S.G.; Zhou, H. MoNbTaV medium-entropy alloy. *Entropy* **2016**, *18*, 189. [CrossRef]
34. Yao, H.W.; Qiao, J.W.; Gao, M.C.; Hawk, J.A.; Ma, S.G.; Zhou, H.F.; Zhang, Y. NbTaV-(Ti,W) refractory high-entropy alloys: Experiments and modeling. *Mater. Sci. Eng. A* **2016**, *674*, 203–211. [CrossRef]
35. *Special Metals IN718 Datasheet*; Special Metals Corporation: New Hartford, NY, USA, 2007; Volume 75.
36. Sengupta, A.; Putatunda, S.K.; Bartosiewicz, L.; Hangas, J.; Nailos, P.J.; Peputapeck, M.; Alberts, F.E. Tensile behavior of a new single crystal nickel-based superalloy (CMSX-4) at room and elevated temperatures. *J. Mater. Eng. Perform.* **1994**, *3*, 664–672. [CrossRef]
37. Basak, A.; Holenarasipura Raghu, S.; Das, S. Microstructures and Microhardness Properties of CMSX-4® Additively Fabricated Through Scanning Laser Epitaxy (SLE). *J. Mater. Eng. Perform.* **2017**, *26*, 5877–5884. [CrossRef]
38. *HAYNES® 282® Alloy—Principal Features*; Haynes International: Kokomo, IN, USA, 2020.

metals

MDPI

Article

Laser Beam Welding of a Low Density Refractory High Entropy Alloy

Evgeniya Panina [1,2,*], Nikita Yurchenko [1], Sergey Zherebtsov [1], Nikita Stepanov [1], Gennady Salishchev [1], Volker Ventzke [2], René Dinse [2] and Nikolai Kashaev [2]

1 Laboratory of Bulk Nanostructured Materials, Belgorod National Research University, Pobeda 85, Belgorod 308015, Russia; yurchenko_nikita@bsu.edu.ru (N.Y.); zherebtsov@bsu.edu.ru (S.Z.); stepanov@bsu.edu.ru (N.S.); salishchev@bsu.edu.ru (G.S.)

2 Department of Laser Processing and Structural Assessment, Institute of Materials Research, Materials Mechanics, Helmholtz-Zentrum Geesthacht, Max-Planck-Str. 1, 21502 Geesthacht, Germany; volker.ventzke@hzg.de (V.V.); rene.dinse@hzg.de (R.D.); nikolai.kashaev@hzg.de (N.K.)

* Correspondence: panina_e@bsu.edu.ru; Tel.: +7-472-258-5416

Received: 12 November 2019; Accepted: 11 December 2019; Published: 16 December 2019

Abstract: The effect of laser beam welding on the structure and properties of a $Ti_{1.89}NbCrV_{0.56}$ refractory high entropy alloy was studied. In particular, the effect of different pre-heating temperatures was examined. Due to the low ductility of the material, laser beam welding at room temperature resulted in the formations of hot cracks. Sound butt joints without cracks were produced using pre-heating to T ≥ 600 °C. In the initial as-cast condition, the alloy consisted of coarse bcc grains with a small amount of lens-shaped C15 Laves phase particles. A columnar microstructure was formed in the welds; the thickness of the grains increased with the temperature of pre-heating before welding. The Laves phase particles were formed in the seams after welding at 600 °C or 800 °C, however, these particles were not observed after welding at room temperature or at 400 °C. Soaking at elevated temperatures did not change the microstructure of the base material considerably, however, "additional" small Laves particles formed at 600 °C or 800 °C. Tensile test of welded specimens performed at 750 °C resulted in the fracture of the base material because of the higher hardness of the welds. The latter can be associated with the bcc grains refinement in the seams.

Keywords: high entropy alloys; laser beam welding; microstructure; mechanical properties

1. Introduction

There is a strong need for the development of novel metallic materials, which would be able to withstand high temperatures for use in critical industries, including aerospace or energy [1]. Meanwhile, suitable commercial alloys (like Ni-based superalloys) do not provide many opportunities for further improvements. Therefore, new alloying concepts must be explored. One of the concepts is associated with so-called high entropy alloys (HEAs)–alloys, composed of several principal components taken in close to equiatomic proportions [2,3]. These alloys were found to possess some unusual properties outperforming conventional materials, which makes HEAs attractive for a variety of applications [3–6].

For example, HEAs, that are composed of refractory elements (refractory high entropy alloys (RHEAs) [7]), were introduced by Senkov and co-workers almost a decade ago [8]. Due to the striking ability to maintain high strengths at extraordinary high temperatures of ~1600 °C [9], RHEAs have instantly attracted considerable attention from the scientific community. First, RHEAs composed of Nb, Mo, Ta, V, and W were too heavy: ~12.0–14.0 g/cm^3 [8,9]. Such a high density made their potential usage in the aerospace industry quite questionable, despite the impressive high-temperature strength. Therefore, significant efforts were undertaken to produce alloys with lower density, but superior high-temperature strength, by using refractory elements with lower specific gravity like Ti and

Cr [10,11] and/or by adding light elements like Al [12,13]. Many new alloys with densities in the range ~5.5–8.0 g/cm^3 (i.e., lighter than the Ni-based superalloys) were introduced; some of them demonstrated highly attractive specific strength at temperatures up to 1200 °C [7,14–21].

High (specific) strength itself does not guarantee the applicability of structural materials, since many other properties are critically important as well [22]. For example, technological properties of RHEAs, which are essential for different products fabrication, remain mostly unexplored. Some alloys have demonstrated reasonable workability at room temperature allowing cold rolling to a high thickness reduction. However, the number of cold-workable RHEAs is quite limited [23–26] so far. Most RHEAs have quite low ductility at room temperature even in compression and thus can be processed at high temperatures only [7].

Weldability is another crucial technological property of the structural materials, since welding is one of the most reliable and efficient ways of joining different parts together. A few studies on welding of HEAs were reported recently [27–32]. The efficiency of using arc welding [28], laser beam welding (LBW) [31], electron beam welding [27,28], and friction stir welding [29,30,32] was shown to obtain sound joints in HEAs with reasonable mechanical properties. However, it should be noted that all of the studies were focused on welding of HEAs composed of non-refractory elements like Co, Fe, Ni, etc. Meanwhile, there is no information in the literature on weldability of RHEAs. Therefore, in the present work we have explored the structure and mechanical properties of the $Ti_{1.89}CrNbV_{0.56}$ RHEA butt joints obtained by LBW. This recently introduced RHEA [33] had a low density of 6.17 g/cm^3 and can be cold rolled to a high thickness reduction. The LBW technique was previously used successfully to join Ti alloys [34–39], and therefore, can be considered a proper welding method for the program Ti-rich RHEA.

The subject of the present work was to determine suitable parameters and process conditions for laser beam welding of the RHEA in order to achieve defect-free butt joints. Furthermore, the microstructure and mechanical properties of the welded joints were characterized. In particular, differences in microstructure between the weld and the base material were of interest.

2. Materials and Methods

Button-shaped ingots of the $Ti_{1.89}CrNbV_{0.56}$ alloy measured ~60 mm in diameter and ~12 mm in height were produced by arc melting in a low-pressure, high-purity argon atmosphere inside a water-cooled copper cavity. The purities of the alloying elements were no less than 99.9 wt. %. The actual chemical composition of the alloy as per energy dispersive X-ray analysis (EDX) measurement with the scan area ~1 mm × 1 mm is presented in Table 1.

Table 1. Chemical composition of the structural constituents of the $Ti_{1.89}CrNbV_{0.56}$ alloy, in at. %.

Structural Constituent	Ti	Cr	Nb	V
Interdendrite	39.8	24.7	24.2	11.3
Alloy	43.7	22.1	22.3	11.9

Specimens for welding measured 40 mm × 15 mm × 2 mm were extracted from the as-cast ingots. Clear shiny surfaces (Ra–1.8 (according to ISO 1997)) on each side of the plates were obtained by mechanical polishing. Butt joint LBW was performed using an 8.0 kW continuous wave ytterbium fiber laser with a fiber optic (300 µm core diameter, 300 mm focal length, 120 mm collimation lens and 750 µm focus diameter). The wavelength of the laser was 1070 nm and the resulting beam parameter product was 10.6 mm × mrad. Welding was performed in horizontal position 2 G: the plane of the specimens was vertical, and the weld seam was horizontal. The weld coupon set-up was mounted on a linear motion unit, which was positioned in a chamber filled with argon protective gas (the Ar flow rate was 40 L/min). A heating device was used to pre-heat the specimens before welding in order to prevent the formation of cracks. Pre-heating temperatures of the specimens were measured by a thermocouple. Details of the LBW set-up can be found in [39].

The LBW process was performed using the following process parameters:

- The laser power of 2.5 kW;
- The focus position of 0.0 mm above the specimen surface;
- The welding speed of 5 m/min;
- The pre-heating temperatures before LBW were 400 °C, 600 °C, and 800 °C; LBW at room temperature was also carried out for the sake of comparison.

The X-ray inspections were used to determine any inner imperfections in the welds, including porosity and cracks. The structure of the alloy was examined using X-ray diffraction (XRD) analysis, scanning (SEM), and transmission (TEM) electron microscopy. Specimens for SEM and XRD were prepared by careful mechanical polishing. XRD analysis was performed using a Rigaku diffractometer (Rigaku Corporation, Tokyo, Japan) with CuKα radiation. SEM back-scattered electron (BSE) images were obtained using FEI Quanta 600 FEG microscope (FEI, Hillsboro, OR, USA) equipped with an EDX detector.

In addition, electron-backscattered diffraction (EBSD) analysis was performed (JEOL JSM-6490LV (JEOL, Akishima, Japan), EDAX TSL OIM (EDAX Inc, Mahwah, NJ, USA)). EBSD measurements were conducted at 30 kV, an emission current of 75 μA, a sample tilt angle of 70°, a working distance of 13 mm and a step size of 2.0 μm. The crystal orientation calculation was based on the generalized spherical harmonic expansion method, where triclinic sample symmetry could be assumed. The linear intercept method was used to measure the size of the grains.

The specimen for TEM analysis were prepared by conventional twin-jet electro-polishing of mechanically pre-thinned to 100 μm foils, in a mixture of methanol (600 mL), butanol (360 mL), and perchloric acid (60 mL) at −35 °C and an applied voltage of 29.5 V. TEM investigations were performed using JEOL JEM-2100 microscope (JEOL, Akishima, Japan) with an accelerating voltage of 200 kV.

Microhardness profiles across the joint were obtained using an automated Vickers hardness testing machine (Instron, Norwood, MA, USA). Nanohardness was determined via the Oliver and Pharr method [40] using Shimadzu DUH-211s Dynamic Ultra Micro Hardness Tester equipped with a Berkovich indenter (both-Shimadzu, Kyoto, Japan). At least ten indents, per structural constituent were performed with the maximum load of 50 mN for 5 s; the loading speed was 6.66 mN/s.

Dog-bone-shaped tensile specimens, with a thickness of 0.5 mm, were cut out using an electric discharge machine from the as-cast and welded specimens. The length and width of the gauge section were 10, and 2 mm, respectively. In the welded specimens, the seam, located in the center of the gage, was perpendicular to the loading direction. Tensile tests were carried out on a 5-kN electro-mechanic universal testing machine (Instron, Norwood, MA, USA) with a constant crosshead speed. The crosshead displacement was measured with a laser extensometer (Instron, Norwood, MA, USA). The tests were performed at 750 °C, due to the low ductility of the alloy. The specimens were held at this temperature for 5 min before the onset of the test. The temperature was controlled by the thermocouple positioned at the center of the specimen gage. Three specimens, in each condition, were tested and characteristic stress-strain curves were shown.

The equilibrium phase diagram was constructed using Thermo-Calc (version 2019b) software and a TCHEA3 database (both–Thermo-Calc AB, Solna, Sweden).

3. Results

3.1. Initial Microstructure

The structure of the $Ti_{1.89}CrNbV_{0.56}$ alloy in the initial as-cast condition consisted of coarse bcc grains with noticeable dendritic segregations (Figure 1a). The average size of the bcc grains was ~280 μm. However, both relatively fine (~100 μm) and very coarse (~500 μm) grains could also be found (Figure 1b). The grains had mostly an irregular shape with curved boundaries. As per chemical analysis the dendritic areas (light ones) were enriched with Ti, while the inter-dendritic space (darker areas) were enriched with Cr and Nb (Table 1). In addition, a small amount (~1%) of lens-shaped second phase particles were found predominantly in the inter-dendritic areas (see the high magnification insert in Figure 1a). According to XRD (Figure 1c) and taking into account previous results for the same alloy [33], these particles can be identified as the C15 (cubic) Laves phase. Note that the presence of the C15 Laves particles with a similar morphology was recently revealed in the $Cr_{10}Nb_{30}Ti_{30}Zr_{30}$ alloy [41]. Small size of the particles (width less than 0.5 μm) did not allow identification of their chemical composition with SEM-EDX system reliably, however these particles were enriched with Cr and Nb.

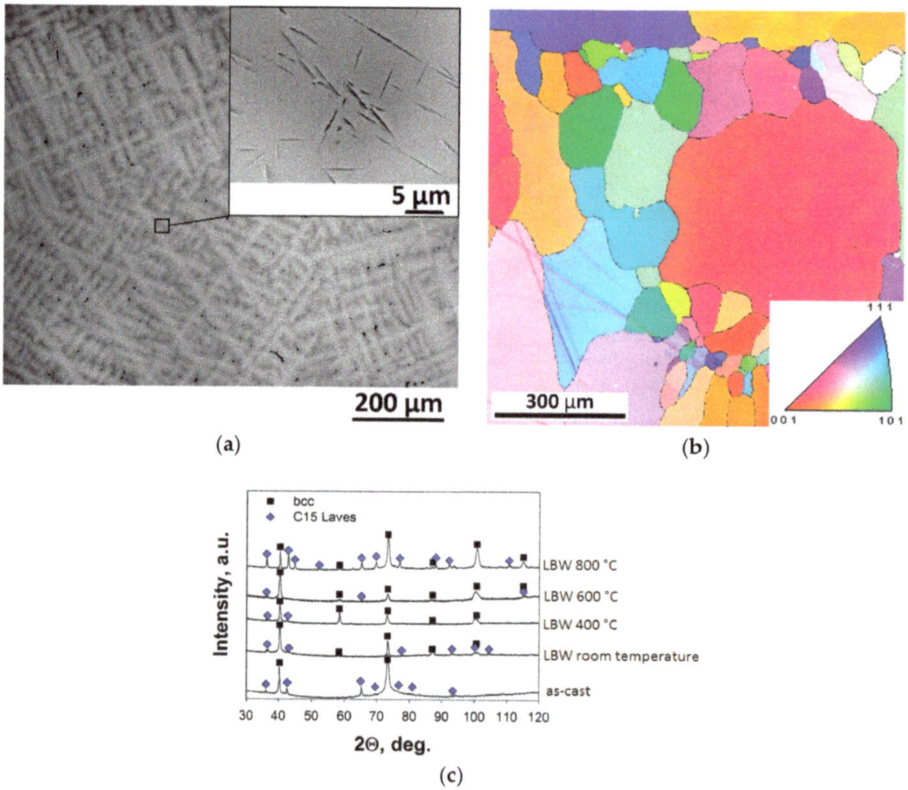

Figure 1. Structure of the $Ti_{1.89}CrNbV_{0.56}$ alloy in the initial as-cast condition: (**a**) SEM-BSE image; (**b**) crystal orientation map; (**c**) XRD pattern.

3.2. Microstructure after Welding

Figure 2 shows the results of X-ray inspections. Well-shaped homogeneous welds were achieved at room and at all pre-heating temperatures. Butt joints welded at room temperature showed several cracks perpendicular to the welding direction (Figure 2a). Specimens welded at 400 °C showed only a single crack in the central part of the specimen (Figure 2b). No cracks were identified after welding at 600 °C or 800 °C (Figure 2c,d). The present results show that the tendency to hot cracking decreases with an increase in the pre-heating temperature before welding.

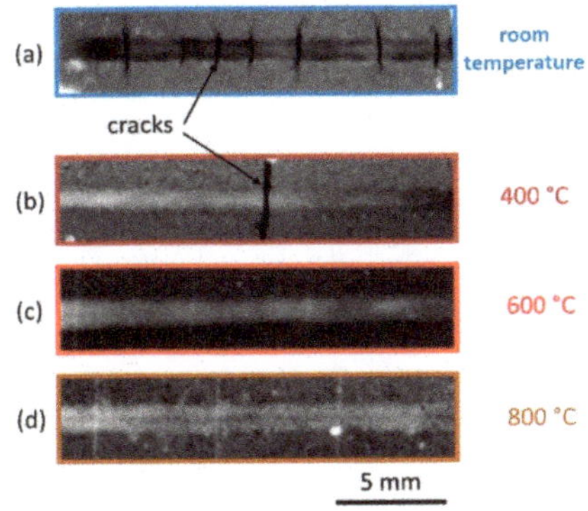

Figure 2. Radiographs of butt joints welded at room temperature (**a**) or at different pre-heating temperatures; (**b**) 400 °C; (**c**) 600 °C, and (**d**) 800 °C.

Grain structure of the welds obtained at different pre-heating temperatures was examined using EBSD technique (Figure 3); the dependence of the transversal size of the bcc grain on the pre-heating temperature is shown in Figure 4. In general, the microstructure of the welds is typical of metallic ingots [42]. After welding at room temperature, a columnar type structure was mainly formed in the weld (Figure 3a). The average width of the grains was ~27 μm (Figure 4); the long axis of the grains was oriented perpendicularly to the joint. At the fusion line between the fusion zone and the heat-affected zone fine (~10 μm), nearly equiaxed grains can be found. These grains were obviously formed at the initial stages of liquid metal solidification due to rapid heat transfer towards the "cold" base metal. Some coarse (~50 μm), nearly equiaxed grains were found at the center of the weld. An increase in the pre-heating temperature to 400 °C resulted in the formation of a more equiaxed structure (Figure 3b) with the average width of the columnar grains of ~40 μm (Figure 4). However, the microstructure had the same morphology typical of metallic ingots: Fine and coarse equiaxed grains at the fusion line between the fusion zone and heat-affected zone, and in the center of the weld, respectively, separated by a columnar grains region. Further increase in the pre-heating temperature during welding resulted in some coarsening of the microstructure (Figure 3c,d). For example, the average width of the columnar grains was ~50 μm or ~70 μm after welding at 600 °C or 800 °C, respectively (Figure 4).

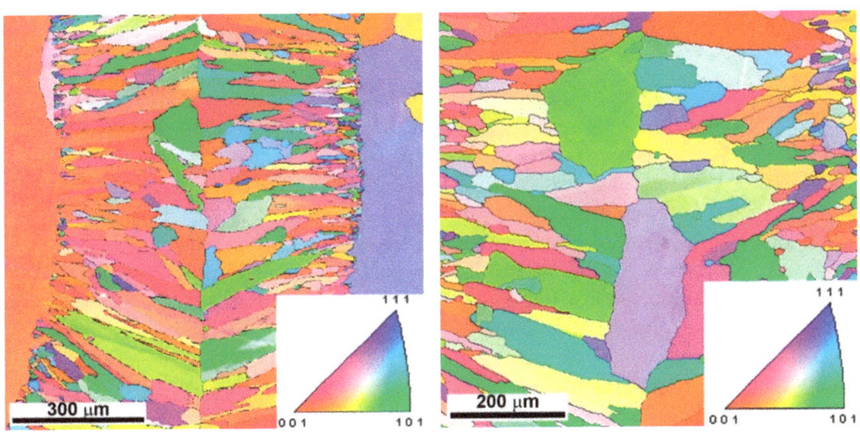

(**a**) (**b**)

Figure 3. *Cont.*

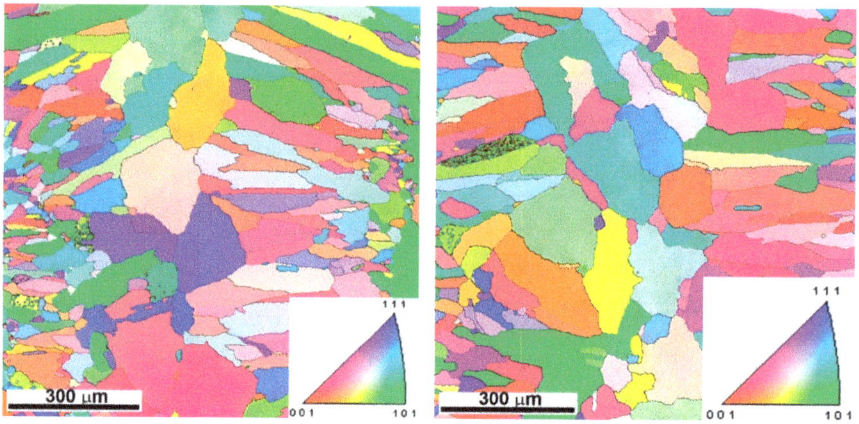

(c) (d)

Figure 3. Crystal orientation maps of butt joint cross-section (seams are vertical in all cases) obtained by laser beam welding (LBW) with different pre-heating temperatures: (**a**) Room temperature; (**b**) 400 °C; (**c**) 600 °C; (**d**) 800 °C.

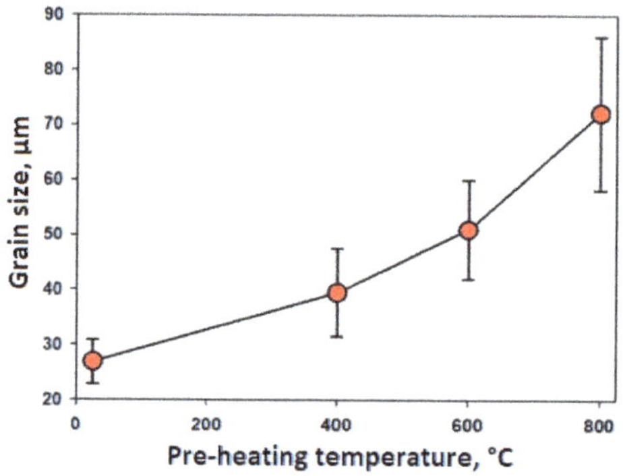

Figure 4. Dependence of the transversal size of the columnar grains inside the welds on the pre-heating temperature.

SEM images of the welding zone are presented in Figure 5. One can see that the dendritic structure survived after welding. However, the width of the dendrite arms and the distance between them gradually increased with the pre-heating temperature, due to lower cooling rates of the melted alloy during solidification. In addition to the changes in the scale of dendrite structure, the pre-heating temperature affected the phase structure of the alloy in the weld. No second phase particles were detected in the samples welded at room temperature or 400 °C (see high magnification inserts in Figure 5a,b). An increase in the pre-heating temperature resulted in the Laves phase particles precipitation predominantly nearby the inter-dendritic areas (Figure 5c,d). The size and fraction of the particles also increased with the pre-heating temperature. Note that the XRD patters showed the presence of the Laves phase in all conditions (Figure 1c). However, these particles were most likely located in the base material (see below).

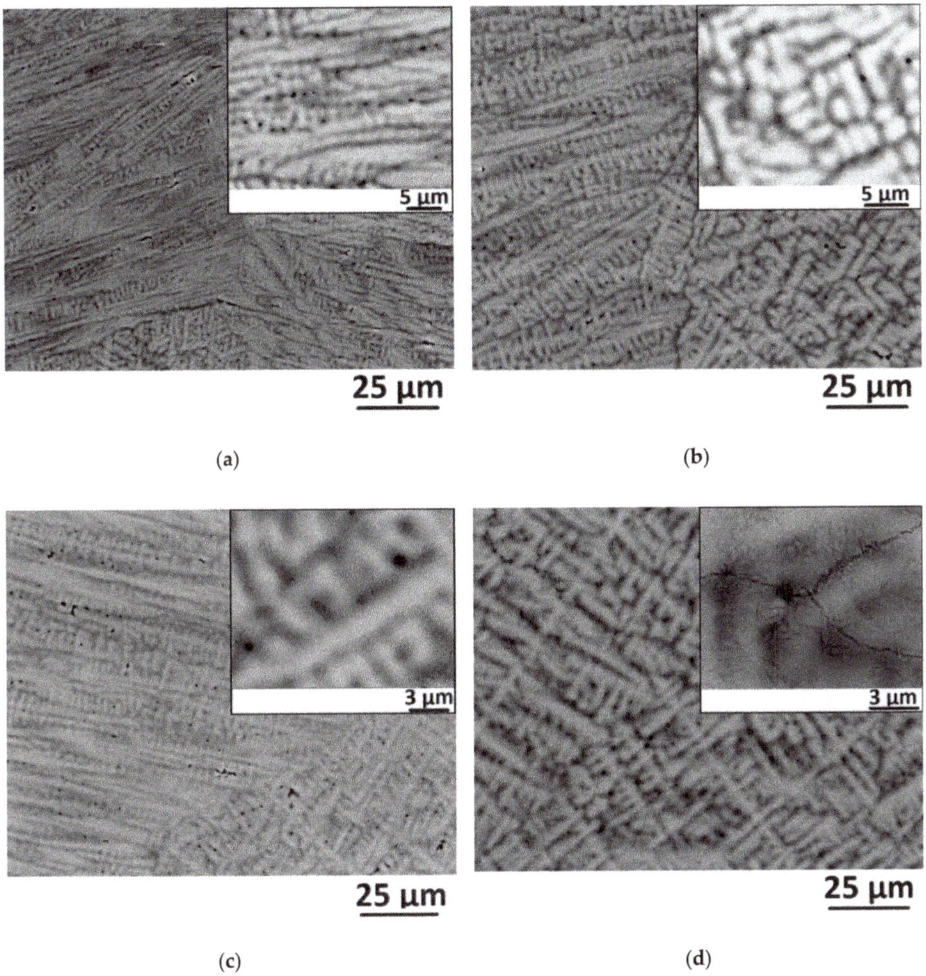

Figure 5. SEM-BSE images of the welding zone after LBW at different pre-heating temperatures: (**a**) room temperature; (**b**) 400 °C; (**c**) 600 °C; (**d**) 800 °C.

Figure 6 shows SEM-BSE images of the base material taken far enough (~5 mm) from the welds. In the specimens welded at T ≤ 400 °C (Figure 6a,b) the microstructure was nearly identical to that in the initial as-cast state (Figure 1a) and was presented by the dendritic segregations with the embedded lens-shaped Laves particles in the bcc coarse-grained matrix. The microstructures of the specimens welded at 600 °C or 800 °C (Figure 6c,d) contained in addition fine dark precipitates located nearby the lens-shaped Laves particles. The fraction of these "additional" particles increased considerably with an increase in the pre-heating temperature from 600 °C to 800 °C. Further TEM investigation revealed that these particles were also in the C15 Laves phase (Figure 7).

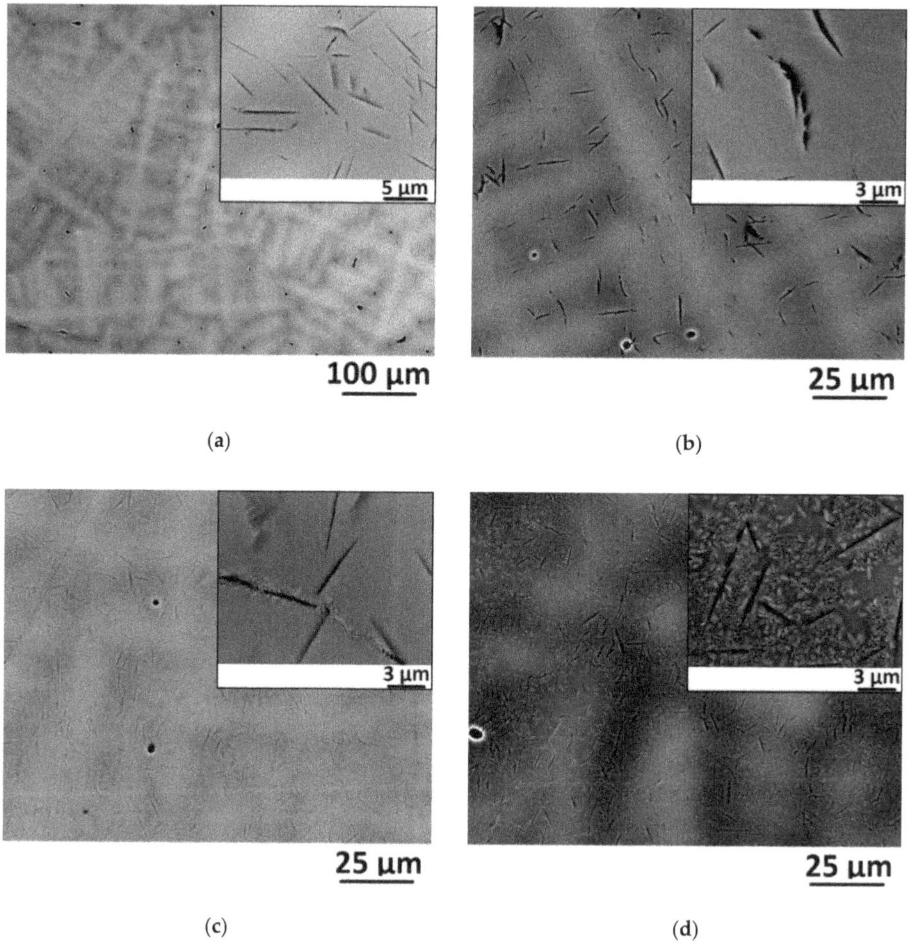

Figure 6. SEM-BSE images of the base material in specimens subjected to LBW at different pre-heating temperatures: (**a**) room temperature; (**b**) 400 °C; (**c**) 600 °C; (**d**) 800 °C.

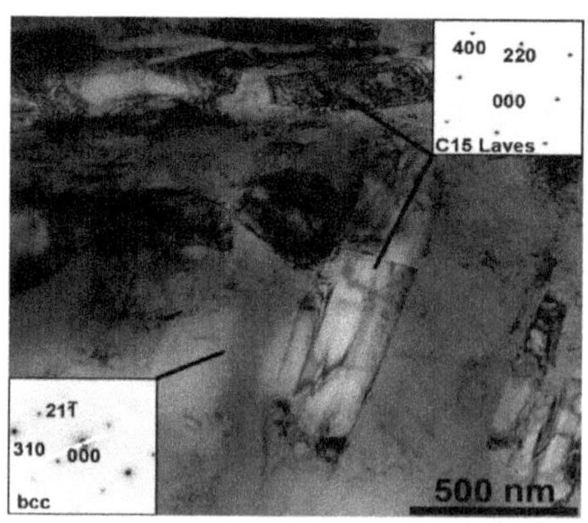

Figure 7. TEM bright-field image of the base material in the specimen after LBW at the pre-heating temperature of 800 °C.

3.3. Mechanical Properties

Figure 8 shows microhardness profiles across the weld obtained at room (Figure 8a) or elevated pre-heating temperatures (Figure 8b). There is no significant variation in the hardness profiles depending on the measurement location (radiation exposure side, half-thickness, or weld root side) after welding at room temperature (Figure 8a). The hardness increases from 475 HV in the base material to 550 HV in the fusion zone after welding at room temperature. An increase in the pre-heating temperature during welding slightly decreased the hardness of the base material until it reached 450 HV after welding at 800 °C. However, the pre-heating temperature had a much more pronounced effect on the fusion zone hardness. The differences between the base material and fusion zone hardness became lower with an increase in the pre-heating temperature, and no hardness gradient can be detected after welding at 800 °C (Figure 8b).

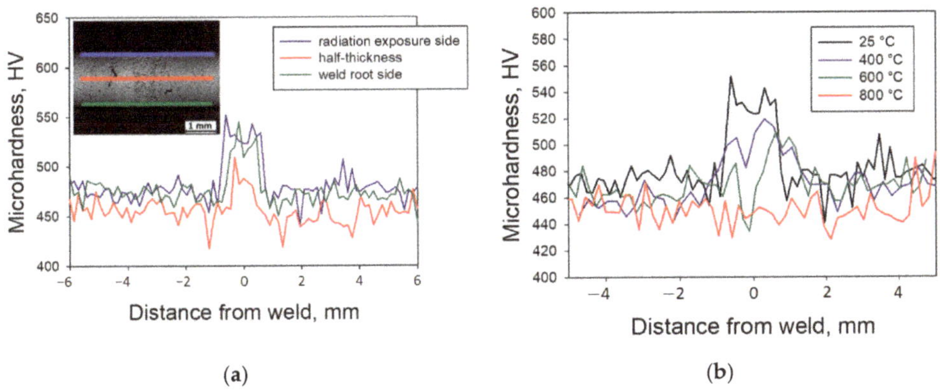

Figure 8. Microhardness profiles across the weld: (**a**) at room temperature along radiation exposure side (blue line), half-thickness (red line) and weld root side (green line); (**b**) at different pre-heating temperature before welding along the half-thickness line.

To get a better understanding of the welding conditions effect on the mechanical properties of the alloy, tensile tests were performed. Tensile stress-strain curves obtained at 750 °C for the as-cast and laser beam welded at 400–800 °C specimens are shown in Figure 9. Note that the specimens welded at room temperature were not tested, since the seams contained numerous cracks (Figure 2a) and therefore could not show suitable properties. The as-cast alloy also fractured in the elastic region at 250 Mpa without any tensile ductility (Table 2). After LBW at 400 °C, the alloy demonstrated almost the same behavior. An increase in the pre-heating temperature before welding to 600 °C resulted in a drastic decrease in the fracture stress to 45 Mpa. However, after 800 °C the alloy showed reasonable ductility of ~10%. The yield strength of this condition was 265 Mpa; after a short hardening stage, and by reaching the ultimate tensile strength of 285 Mpa, an extensive softening stage associated with strain localization was observed. In all cases, fractures occurred in the base material.

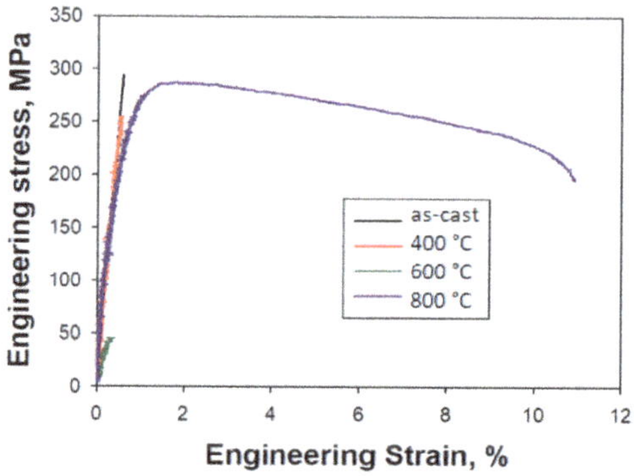

Figure 9. Engineering stress-strain curves of the $Ti_{1.89}NbCrV_{0.56}$ alloy tensioned at 750 °C in the as-cast state and after LBW at 400, 600 or 800 °C.

Table 2. Mechanical properties (YS—yield strength, UTS—ultimate tensile strength, FS—fracture stress, δ—elongation to fracture) of the $Ti_{1.89}NbCrV_{0.56}$ alloy obtained by tensile tests at 750 °C in the as-cast state and after LBW at 400, 600 or 800 °C.

Condition	YS, Mpa	UTS (or FS), Mpa	δ, %
As-cast	-	(250)	0
LBW 400 °C	-	(255)	0
LBW 600 °C	-	(45)	0
LBW 800 °C	265	285	10.4

Fracture surfaces of laser beam welded specimens at 400 °C or 800 °C after tensile tests were quite different (Figure 10). At the lower pre-heating temperature during welding, a typical brittle fracture with cleavage planes and river pattern on some facets, was observed. Secondary cracks were formed along some grain boundaries. At the higher pre-heating temperature (800 °C) the character of fracture became more ductile; the fracture surface was consisted of relatively small (~5 μm) dimples due to void coalescence.

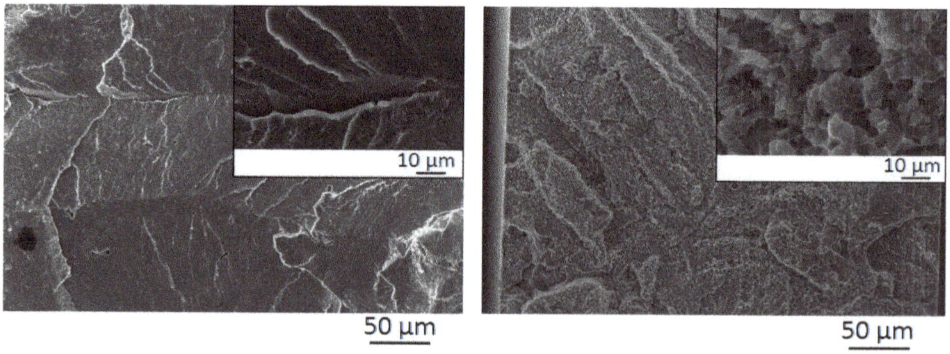

(a) (b)

Figure 10. Fracture surfaces of the tensile specimens after LBW at different pre-heating temperatures: (**a**) 400 °C; (**b**) 800 °C.

4. Discussion

In this work, a butt joint of the $Ti_{1.89}CrNbV_{0.56}$ RHEA, using laser beam welding, was successfully produced for the first time. Due to a high melting temperature and brittleness of the material, LBW was performed at different pre-heating temperatures from room temperature to 800 °C to prevent the formation of hot cracks. Apparently, the pre-heating temperature affected both the microstructure and mechanical properties of the welded specimens significantly.

The initial, as-cast microstructure was quite heterogeneous and consisted of coarse bcc matrix grains with the lens-shaped Laves phase particles (Figure 1). LBW has obviously affected the morphology of the bcc grains in the welding zone (WZ) (Figure 3). The grains became elongated in the direction normal to the laser beam, and the width of these grains increased with pre-heating temperature, most likely due to enhanced mobility of grain boundaries at higher pre-heating temperatures. In addition, LBW at the lower pre-heating temperatures (room temperature and 400 °C) has resulted in the Laves phase particles disappearance (Figure 5a,b). The formation of these particles in the alloy deserves a more detailed analysis. The equilibrium phase diagram for the alloy constructed by the Thermo-Calc software is presented in Figure 11.

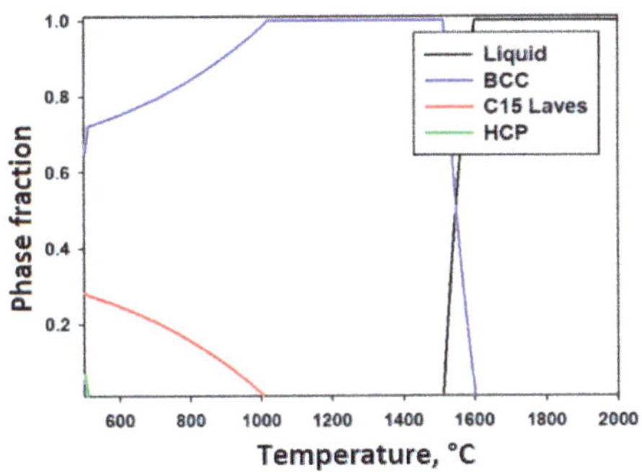

Figure 11. Phase diagram of the $Ti_{1.89}NbCrV_{0.56}$ alloy constructed using the Thermo-Calc software (BCC—body-centered cubic structure; HCP—hexagonal close packed structure).

The alloy is supposed to have a single bcc phase structure starting from the solidus temperature of ~1500 °C till ~1000 °C; the latter corresponds to the onset of the Cr-rich C15 Laves phase precipitation. The fraction of the Laves phase increased with a decrease in temperature. Finally, at T ≈ 500 °C a Ti-rich hcp phase precipitated. The presence of the Laves phase in the initial as-cast condition can be most probably associated with a rather low cooling rate of relatively big ingots of the alloy that was not enough to "freeze" the high-temperature single-phase structure. In turn, faster cooling rates of the alloy, during the welded material solidification, prevented the formation of the Laves particles in the case of LBW at room temperature or 400 °C. During welding at higher pre-heating temperatures, the cooling rate is slower thereby allowing the Laves particles precipitation.

Also, annealing at 800 °C resulted in the precipitation of "additional" Laves particles in the base material after welding at pre-heating temperature of 600 °C or 800 °C. The formation of the particles can be attributed to the decomposition of the supersaturated bcc solid solution during holding at elevated temperatures (Figure 11). Note that the particles precipitated in the inter-dendritic areas were observed mainly nearby the coarser particles, inherited from the as-cast condition. The Laves phase in this alloy is supposed to be composed of Cr and Nb predominantly [33] and, therefore, the "additional" particles formed in the inter-dendritic areas which were rich in these elements (Table 1), similarly to the Laves phase particles formation in the $NbCrMo_{0.5}Ta_{0.5}TiZr$ alloy [43].

Mechanical properties of the $Ti_{1.89}NbCrV_{0.56}$ alloy were found to be strongly sensitive to the welding conditions. A noticeable increase in the fusion zone hardness, as well as the variation in the hardness with the temperature during welding (Figure 8b), can be attributed to the bcc matrix grain size as a function of pre-heating temperature (Figure 4). The relationship between the average hardness and average grain size in the fusion zone shown in Figure 12 suggests that the Hall-Petch mechanism is active.

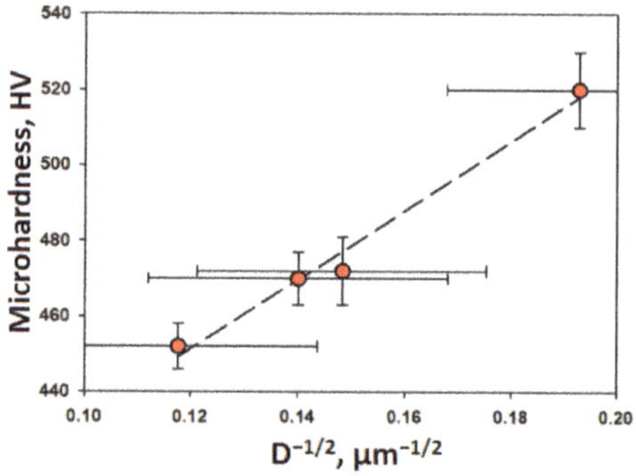

Figure 12. Relationship between the hardness and grain size in the fusion zone.

Due to the higher hardness of the weld, welded specimens fractured in the base material section during tensile tests. This type of behavior was observed earlier in other welded HEAs [44]. However, the welding conditions strongly influenced on tensile behavior; for example, specimens welded at T ≤ 600 °C exhibited brittle fracture even at 750 °C while the alloy welded at 800 °C showed reasonable ductility (Figure 9, Table 2). One of the reasons for the increased ductility can be associated with the formation of a less defect structure (in terms of cracks) with an increase in the pre-heating temperature before welding (Figure 2).

The "additional" Laves particles precipitation after welding at 800 °C can also provoke an increase in ductility. It is generally believed that the second phase particles precipitation should result in strengthening. However, in the present case nano-hardness of the particles-containing areas (480 HV) was found to be lower than that of the particle-free zone (560 HV) after welding at 800 °C. This result can most likely be related to (i) lower solid solution hardening of the bcc matrix due to depletion in Cr and Nb [11,33] and (ii) low (er) hardness of the Laves phase itself [33]. That is why the Laves phase particles precipitation in the present alloy resulted in both some softening and an increase in ductility after welding at 800 °C. It should be also noted, that the size of the facets in Figure 10b is correlated with the size of precipitation-free areas (Figure 5d) thereby suggesting that the more ductile behavior can be associated with a temporary crack arrest in these Cr and Nb depleted regions.

The obtained results demonstrate for the first time the possibility to produce sound butt joints of the $Ti_{1.89}NbCrV_{0.56}$ RHEA by LBW, which is an important step toward potential practical applications of this new class of metallic alloys. However, the present study also emphasized the importance of LBW process parameters (particularly pre-heating temperature) to obtain good results, which is quite different from the transition metals HEAs [27–32,44,45]. Given the wide range of the available RHEAs compositions [7], specific attention should be paid to select proper welding conditions for each individual alloy.

5. Conclusions

In this work, laser beam welding, at different pre-heating temperatures, was successfully used to produce butt joints of the $Ti_{1.89}NbCrV_{0.56}$ refractory high entropy alloy. The following conclusions were drawn:

1. The as-cast $Ti_{1.89}NbCrV_{0.56}$ alloy had a coarse-grained bcc structure with a small amount of lens-shaped C15 Laves phase particles. Welding at room temperature or using the pre-heating temperature of 400 °C did not result in the formation of the Laves phase particles in the seams. However, these particles were found after welding with the pre-heating temperature of 600 °C or 800 °C. Newly crystallized bcc grains in the welds were mostly elongated; their thickness increased with an increase in the pre-heating temperature. The structure of the base material did not change noticeably after welding except for the "additional" Laves particles precipitation

after soaking at 600 or 800 °C during LBW. Cracks were observed in the weld after LBW at room temperature or at 400 °C.
2. Specimens welded at the pre-heating temperature of 800 °C demonstrated reasonable tensile ductility at 750 °C; welding at the lower pre-heating temperatures resulted in brittle fracture during tension. Fracture in all cases occurred in the base material. The hardness of the weld was higher than that of the base material due to the bcc grains refinement.

Author Contributions: Conceptualization, S.Z., N.S., N.K.; methodology, N.K., G.S., V.V.; validation, E.P., N.K.; formal analysis, N.K., N.S.; investigation, E.P., N.Y., V.V., R.D.; resources, G.S., N.K.; data curation, E.P.; writing—original draft preparation, N.S.; writing—review and editing, S.Z., N.K.; visualization, E.P., N.Y.; supervision, N.K., G.S.; project administration, S.Z.; funding acquisition, G.S.

Funding: This research was funded by the Russian Science Foundation Grant no. 19-79-30066.

Acknowledgments: The authors are grateful to the personnel of the Joint Research Center, Technology and Materials, Belgorod National Research University, for their assistance with the instrumental analysis.

Conflicts of Interest: The authors declare no conflict of interest.

References

1. Perepezko, J.H. Materials science. The hotter the engine, the better. *Science* **2009**, *326*, 1068–1069. [CrossRef] [PubMed]
2. Yeh, J.W.; Chen, S.K.; Lin, S.J.; Gan, J.Y.; Chin, T.S.; Shun, T.T.; Tsau, C.H.; Chang, S.Y. Nanostructured high-entropy alloys with multiple principal elements: Novel alloy design concepts and outcomes. *Adv. Eng. Mater.* **2004**, *6*, 299–303. [CrossRef]
3. Miracle, D.B.; Senkov, O.N. A critical review of high entropy alloys and related concepts. *Acta Mater.* **2017**, *122*, 448–511. [CrossRef]
4. Gorsse, S.; Miracle, D.B.; Senkov, O.N. Mapping the world of complex concentrated alloys. *Acta Mater.* **2017**, *135*, 177–187. [CrossRef]
5. Gludovatz, B.; Hohenwarter, A.; Catoor, D.; Chang, E.H.; George, E.P.; Ritchie, R.O. A fracture-resistant high-entropy alloy for cryogenic applications. *Science* **2014**, *345*, 1153–1158. [CrossRef]
6. Li, Z.; Pradeep, K.G.; Deng, Y.; Raabe, D.; Tasan, C.C. Metastable high-entropy dual-phase alloys overcome the strength-ductility trade-off. *Nature* **2016**, *534*, 227–230. [CrossRef]
7. Senkov, O.N.; Miracle, D.B.; Chaput, K.J.; Couzinie, J.-P. Development and exploration of refractory high entropy alloys—A review. *J. Mater. Res.* **2018**, *33*, 3092–3128. [CrossRef]
8. Senkov, O.N.; Wilks, G.B.; Miracle, D.B.; Chuang, C.P.; Liaw, P.K. Refractory high-entropy alloys. *Intermetallics* **2010**, *18*, 1758–1765. [CrossRef]
9. Senkov, O.N.; Wilks, G.B.; Scott, J.M.; Miracle, D.B. Mechanical properties of $Nb_{25}Mo_{25}Ta_{25}W_{25}$ and $V_{20}Nb_{20}Mo_{20}Ta_{20}W_{20}$ refractory high entropy alloys. *Intermetallics* **2011**, *19*, 698–706. [CrossRef]
10. Senkov, O.N.; Senkova, S.V.; Woodward, C.; Miracle, D.B. Low-density, refractory multi-principal element alloys of the Cr-Nb-Ti-V-Zr system: Microstructure and phase analysis. *Acta Mater.* **2013**, *61*, 1545–1557. [CrossRef]
11. Senkov, O.N.; Senkova, S.V.; Miracle, D.B.; Woodward, C. Mechanical properties of low-density, refractory multi-principal element alloys of the Cr-Nb-Ti-V-Zr system. *Mater. Sci. Eng. A* **2013**, *565*, 51–62. [CrossRef]
12. Senkov, O.N.; Senkova, S.V.; Woodward, C.F. Effect of aluminum on the microstructure and properties of two refractory high-entropy alloys. *Acta Mater.* **2014**, *68*, 214–228. [CrossRef]
13. Senkov, O.N.; Woodward, C.; Miracle, D.B. Microstructure and properties of aluminum-containing refractory high-entropy alloys. *JOM* **2014**, *66*, 2030–2042. [CrossRef]
14. Stepanov, N.D.; Shaysultanov, D.G.; Salishchev, G.A.; Tikhonovsky, M.A. Structure and mechanical properties of a light-weight AlNbTiV high entropy alloy. *Mater. Lett.* **2015**, *142*, 153–155. [CrossRef]
15. Stepanov, N.D.; Yurchenko, N.Y.; Skibin, D.V.; Tikhonovsky, M.A.; Salishchev, G.A. Structure and mechanical properties of the AlCrxNbTiV (x = 0, 0.5, 1, 1.5) high entropy alloys. *J. Alloys Compd.* **2015**, *652*, 266–280. [CrossRef]
16. Stepanov, N.D.; Yurchenko, N.Y.; Panina, E.S.; Tikhonovsky, M.A.; Zherebtsov, S.V. Precipitation-strengthened refractory Al0.5CrNbTi2V0.5 high entropy alloy. *Mater. Lett.* **2017**, *188*, 162–164. [CrossRef]

17. Yurchenko, N.Y.; Stepanov, N.D.; Zherebtsov, S.V.; Tikhonovsky, M.A.; Salishchev, G.A. Structure and mechanical properties of B2 ordered refractory $AlNbTiVZr_x$ (x = 0–1.5) high-entropy alloys. *Mater. Sci. Eng. A* **2017**, *704*, 82–90. [CrossRef]
18. Senkov, O.N.; Jensen, J.K.; Pilchak, A.L.; Miracle, D.B.; Fraser, H.L. Compositional variation effects on the microstructure and properties of a refractory high-entropy superalloy AlMo0.5NbTa0.5TiZr. *Mater. Des.* **2018**, *139*, 498–511. [CrossRef]
19. Senkov, O.N.; Rao, S.; Chaput, K.J.; Woodward, C. Compositional effect on microstructure and properties of NbTiZr-based complex concentrated alloys. *Acta Mater.* **2018**, *151*, 201–215. [CrossRef]
20. Chen, H.; Kauffmann, A.; Gorr, B.; Schliephake, D.; Seemüller, C.; Wagner, J.N.; Christ, H.-J.; Heilmaier, M. Microstructure and mechanical properties at elevated temperatures of a new Al-containing refractory high-entropy alloy Nb-Mo-Cr-Ti-Al. *J. Alloys Compd.* **2016**, *661*, 206–215. [CrossRef]
21. Xu, Z.Q.; Cheng, X.W.; Wang, M.; Chen, Y.W.; Tan, Y.D.; Ma, Z.L. Design of novel low-density refractory high entropy alloys for high temperature applications. *Mater. Sci. Eng. A* **2019**, *755*, 318–322. [CrossRef]
22. Miracle, D.; Majumdar, B.; Wertz, K.; Gorsse, S. New strategies and tests to accelerate discovery and development of multi-principal element structural alloys. *Scr. Mater.* **2017**, *127*, 195–200. [CrossRef]
23. Senkov, O.N.; Semiatin, S.L. Microstructure and properties of a refractory high-entropy alloy after cold working. *J. Alloys Compd.* **2015**, *649*, 1110–1123. [CrossRef]
24. Senkov, O.N.; Pilchak, A.L.; Semiatin, S.L. Effect of cold deformation and annealing on the microstructure and tensile properties of a HfNbTaTiZr refractory high entropy alloy. *Metall. Mater. Trans. A* **2018**, *49*, 2876–2892. [CrossRef]
25. Lilensten, L.; Couzinié, J.-P.; Bourgon, J.; Perrière, L.; Dirras, G.; Prima, F.; Guillot, I. Design and tensile properties of a bcc Ti-rich high-entropy alloy with transformation-induced plasticity. *Mater. Res. Lett.* **2017**, *5*, 110–116. [CrossRef]
26. Sheikh, S.; Shafeie, S.; Hu, Q.; Ahlström, J.; Persson, C.; Veselý, J.; Zýka, J.; Klement, U.; Guo, S. Alloy design for intrinsically ductile refractory high-entropy alloys. *J. Appl. Phys.* **2016**, *120*, 164902. [CrossRef]
27. Wu, Z.; David, S.A.; Feng, Z.; Bei, H. Weldability of a high entropy CrMnFeCoNi alloy. *Scr. Mater.* **2016**, *124*, 81–85. [CrossRef]
28. Wu, Z.; David, S.A.; Leonard, D.N.; Feng, Z.; Bei, H. Microstructures and mechanical properties of a welded CoCrFeMnNi high-entropy alloy. *Sci. Technol. Weld. Join.* **2018**, *23*, 585–595. [CrossRef]
29. Zhu, Z.G.; Sun, Y.F.; Ng, F.L.; Goh, M.H.; Liaw, P.K.; Fujii, H.; Nguyen, Q.B.; Xu, Y.; Shek, C.H.; Nai, S.M.L.; et al. Friction-stir welding of a ductile high entropy alloy: Microstructural evolution and weld strength. *Mater. Sci. Eng. A* **2018**, *711*, 524–532. [CrossRef]
30. Zhu, Z.G.; Sun, Y.F.; Goh, M.H.; Ng, F.L.; Nguyen, Q.B.; Fujii, H.; Nai, S.M.L.; Wei, J.; Shek, C.H. Friction stir welding of a CoCrFeNiAl0.3 high entropy alloy. *Mater. Lett.* **2017**, *205*, 142–144. [CrossRef]
31. Kashaev, N.; Ventzke, V.; Stepanov, N.; Shaysultanov, D.; Sanin, V.; Zherebtsov, S. Laser beam welding of a CoCrFeNiMn-type high entropy alloy produced by self-propagating high-temperature synthesis. *Intermetallics* **2018**, *96*, 63–71. [CrossRef]
32. Shaysultanov, D.; Stepanov, N.; Malopheyev, S.; Vysotskiy, I.; Sanin, V.; Mironov, S.; Kaibyshev, R.; Salishchev, G.; Zherebtsov, S. Friction stir welding of a carbon-doped CoCrFeNiMn high-entropy alloy. *Mater. Charact.* **2018**, *145*, 353–361. [CrossRef]
33. Yurchenko, N.Y.; Panina, E.S.; Zherebtsov, S.V.; Tikhonovsky, M.A.; Salishchev, G.A.; Stepanov, N.D. Microstructure evolution of a novel low-density Ti-Cr-Nb-V refractory high entropy alloy during cold rolling and subsequent annealing. *Mater. Charact.* **2019**, *158*, 109980. [CrossRef]
34. Auwal, S.T.; Ramesh, S.; Yusof, F.; Manladan, S.M. A review on laser beam welding of titanium alloys. *Int. J. Adv. Manuf. Technol.* **2018**, *97*, 1071–1098. [CrossRef]
35. Akman, E.; Demir, A.; Canel, T.; Sınmazçelik, T. Laser welding of Ti6Al4V titanium alloys. *J. Mater. Process. Technol.* **2009**, *209*, 3705–3713. [CrossRef]
36. Li, Z.; Gobbi, S.L.; Norris, I.; Zolotovsky, S.; Richter, K.H. Laser welding techniques for titanium alloy sheet. *J. Mater. Process. Technol.* **1997**, *65*, 203–208. [CrossRef]
37. Liu, J.; Dahmen, M.; Ventzke, V.; Kashaev, N.; Poprawe, R. The effect of heat treatment on crack control and grain refinement in laser beam welded β-solidifying TiAl-based alloy. *Intermetallics* **2013**, *40*, 65–70. [CrossRef]

38. Kashaev, N.; Ventzke, V.; Fomichev, V.; Fomin, F.; Riekehr, S. Effect of Nd:YAG laser beam welding on weld morphology and mechanical properties of Ti-6Al-4V butt joints and T-joints. *Opt. Lasers Eng.* **2016**, *86*, 172–180. [CrossRef]
39. Burkhardt, I.; Ventzke, V.; Riekehr, S.; Kashaev, N.; Enz, J. Laser welding and microstructural characterization of dissimilar γ-TiAl-Ti6242 joints. *Intermetallics* **2019**, *104*, 74–83. [CrossRef]
40. Oliver, W.C.; Pharr, G.M. Measurement of hardness and elastic modulus by instrumented indentation: Advances in understanding and refinements to methodology. *J. Mater. Res.* **2004**, *19*, 3–20. [CrossRef]
41. Senkov, O.N.; Zhang, C.; Pilchak, A.L.; Payton, E.J.; Woodward, C.; Zhang, F. CALPHAD-aided development of quaternary multi-principal element refractory alloys based on NbTiZr. *J. Alloys Compd.* **2019**, *783*, 729–742. [CrossRef]
42. Cahn, R.W.; Haasen, P. *Physical Metallurgy*; North-Holland Physics Pub.: Amsterdam, The Netherlands, 1983; ISBN 0444867872.
43. Senkov, O.N.; Woodward, C.F. Microstructure and properties of a refractory NbCrMo0.5Ta0.5TiZr alloy. *Mater. Sci. Eng. A* **2011**, *529*, 311–320. [CrossRef]
44. Kashaev, N.; Ventzke, V.; Petrov, N.; Horstmann, M.; Zherebtsov, S.; Shaysultanov, D.; Sanin, V.; Stepanov, N. Fatigue behaviour of a laser beam welded CoCrFeNiMn-type high entropy alloy. *Mater. Sci. Eng. A* **2019**, *766*, 138358. [CrossRef]
45. Martin, A.C.; Fink, C. Initial weldability study on Al0.5CrCoCu0.1FeNi high-entropy alloy. *Weld. World* **2019**, *63*, 739–750. [CrossRef]

metals

Article

Compositional Design of Soft Magnetic High Entropy Alloys by Minimizing Magnetostriction Coefficient in $(Fe_{0.3}Co_{0.5}Ni_{0.2})_{100-x}(Al_{1/3}Si_{2/3})_x$ System

Yong Zhang [1,*], Min Zhang [1], Dongyue Li [1], Tingting Zuo [2], Kaixuan Zhou [3], Michael C. Gao [4,5], Baoru Sun [3] and Tongde Shen [3]

1 The State Key Laboratory of Advanced Metals and Materials, University of Science and Technology Beijing, Xueyuan Road 30, Beijing 100083, China; zhangminw925@163.com (M.Z.); lidongyue03@163.com (D.L.)
2 Interdisciplinary Research Center, Institute of Electrical Engineering, Chinese Academy of Sciences, Beijing 100190, China; zuott1986.520@163.com
3 State Key Laboratory of Metastable Materials Science and Technology, Yanshan University, Qinhuangdao 066004, China; zkx@stumail.ysu.edu.cn (K.Z.); sunbaoru@ysu.edu.cn (B.S.); tdshen@ysu.edu.cn (T.S.)
4 National Energy Technology Laboratory, Albany, OR 97321, USA; michael.gao@netl.doe.gov
5 Leidos Research Support Team, P.O. Box 10940, Pittsburgh, PA 15236, USA
* Correspondence: drzhangy@ustb.edu.cn; Tel.: +86-186-0008-0235

Received: 31 January 2019; Accepted: 20 March 2019; Published: 26 March 2019

Abstract: Developing cost-effective soft magnetic alloys with excellent mechanical properties is very important to energy-saving industries. This study investigated the magnetic and mechanical properties of a series of $(Fe_{0.3}Co_{0.5}Ni_{0.2})_{100-x}(Al_{1/3}Si_{2/3})_x$ high-entropy alloys (HEAs) (x = 0, 5, 10, 15, and 25) at room temperature. The $Fe_{0.3}Co_{0.5}Ni_{0.2}$ base alloy composition was chosen since it has very the smallest saturation magnetostriction coefficient. It was found that the $(Fe_{0.3}Co_{0.5}Ni_{0.2})_{95}(Al_{1/3}Si_{2/3})_5$ alloy maintains a simple face-centered cubic (FCC) solid solution structure in the states of as-cast, cold-rolled, and after annealing at 1000 °C. The alloy after annealing exhibits a tensile yield strength of 235 MPa, ultimate tensile strength of 572 MPa, an elongation of 38%, a saturation magnetization (Ms) of 1.49 T, and a coercivity of 96 A/m. The alloy not only demonstrates an optimal combination of soft magnetic and mechanical properties, it also shows advantages of easy fabrication and processing and high thermal stability over silicon steel and amorphous soft magnetic materials. Therefore, the alloy of $(Fe_{0.3}Co_{0.5}Ni_{0.2})_{95}(Al_{1/3}Si_{2/3})_5$ holds good potential as next-generation soft magnets for wide-range industrial applications.

Keywords: high-entropy alloy; soft magnetic properties; mechanical properties; saturation magnetostriction coefficient; face-centered cubic (FCC) structure

1. Introduction

Contrasted to the traditional alloy design concept that is based on one or two major elements, high-entropy alloys (HEAs) usually contain four or more major elements, and the content of each major element is above 5 at% [1–10]. The high configurational entropy of the HEAs tends to promote forming the solid solution phases with the face-centered cubic (FCC) [11–15], body-centered cubic (BCC) [16,17], and hexagonal-close-packed (HCP) structures [18–20], as well as dual solid-solution phases [21–23]. The composition of HEAs is generally near the center of the multi-component phase diagram, and hence it is difficult to distinguish the solvent from the solute. The following 5 characteristics are typically and have been verified for the HEAs [4,24,25]:

(1). high thermal stability and resistance to heat softening;
(2). easier to break the tradeoff between strength and ductility;
(3). very low stacking-fault energy;
(4). high irradiation resistance;
(5). high corrosion resistance.

Consequently, HEAs show promising potentials for wide ranges of industrial applications and the HEA concept is considered to be one of the three major breakthroughs in alloying theory [3].

To date considerable research efforts on HEAs focus on their microstructures and mechanical properties [26–31] while there is rising interest in developing high-entropy soft magnetic materials [32–40]. Magnetic materials play a fundamental role in transformers, motors, electromagnets, and other power industries [41]. Common soft magnetic alloys include silicon steel, ferrite, iron-nickel alloys, and bulk metallic glasses. However, to date many challenges still remain for energy-saving and high-performance applications. For example, it is time-consuming to produce and process silicon-steels [42]. Ferrite has low magnetization, iron-nickel alloy has low resistivity, and bulk metallic glass are inherently brittle and are also limited to small size [43,44]. Recent studies have shown that FeCoNiAlSi-based and FeCoNiMnAl-based HEAs exhibit high saturation magnetization which open up new research directions of soft magnetic materials [45]. However, the coercivity of HEAs is a bit high, for example, the coercivity of FeCoNi(Al,Si)$_{0.2}$ alloy is 1400 A/m [46]. It is known that the magnetic properties of HEAs are sensitive to the alloy system and the composition, the phase crystal structures, and the microstructure.

In terms of soft magnetic materials, the most important characteristic is high permeability, high saturation induction, low core loss and low coercivity. In this study, ferromagnetic elements Fe, Co, and Ni were selected as the main constituents of the alloy system [47]. The base alloy composition $Fe_{0.3}Co_{0.5}Ni_{0.2}$ was chosen since the alloy has a close-to-zero saturation magnetostriction coefficient (λs) as shown in the Co-Fe-Ni ternary system (Figure 1). To balance soft magnetic and mechanical properties, Al and Si in small amount was added to $Fe_{0.3}Co_{0.5}Ni_{0.2}$, and the resulting materials properties of the $(Fe_{0.3}Co_{0.5}Ni_{0.2})_{100-X}(Al_{1/3}Si_{2/3})_X$ (X = 0, 5, 10, 15, and 25) high entropy alloys were studied systematically. It was found that the $(Fe_{0.3}Co_{0.5}Ni_{0.2})_{95}(Al_{1/3}Si_{2/3})_5$ alloy exhibits excellent soft magnetic properties. The alloy has small coercivity of 96 A/m, and the saturation magnetic induction (Bs) and the saturation magnetization (Ms) reached 1.55 T and 1.49 T, respectively. The alloy in the as-cast state also shows an excellent tensile elongation up to 25% and a yield stress of 122 MPa at room temperature.

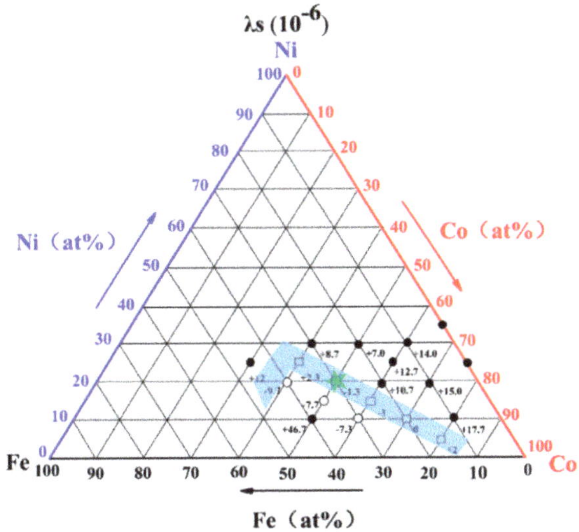

Figure 1. The saturation magnetostriction (λs) of the Co-Fe-Ni bulk ternary alloys at room temperature.

2. Materials and Methods

2.1. Alloys Preparation

Al, Co, Cr, Fe, and Ni elements with purities of greater than 99.5 wt.% were used to prepare $(Fe_{0.3}Co_{0.5}Ni_{0.2})_{100-X}(Al_{1/3}Si_{2/3})_X$ alloys ingots using vacuum arc melting (VAM) and magnetic levitation melting (MLM). The alloys were remelted five times in order to improve the chemical homogeneity. The ingot of the maglev melted alloy is cut into a plate with a thickness of 6 mm. Then the plate was cold rolled to a thickness of 1 mm (i.e., thickness reduction of 83%) by multi-pass rolling, with 0.5 mm reduction for each step. Subsequent annealing heat treatment was performed at 1000 °C for 2 h followed by water quenching or furnace cooling.

2.2. Structural Characterization and Performance Experiments

The crystal structures of the ingots were measured by a Smart Lab X-ray diffractometer (Bruker D8, Karlsruhe, Germany), using Cu-Kα radiation. The electron backscatter diffraction (EBSD) was carried out by a field emission scanning electron microscope (ZEISS SUPRA 55, Jena, Germany). Room temperature compression and tensile tests were conducted on a CMT Model 4305 Universal Electronic Tester with a strain rate of 1×10^{-3} min^{-1}. A hysteresis loop and hysteresis loss curve were obtained by PPMS (physical property measurement system) and VSM (Vibrating Sample Magnetometer) at room temperature. The four-point method was used for measuring the electrical resistance.

3. Results

Figure 2 shows the XRD profiles of $(Fe_{0.3}Co_{0.5}Ni_{0.2})_{100-X}(Al_{1/3}Si_{2/3})_X$ HEAs, (X = 0, 5, 10, 15, and 25). When X = 0, 5, and 10, the alloys possess a simple FCC structure. When X reaches 15, BCC phase appears and the alloy exhibits an FCC + BCC dual-phase structure. When the $(Al_{1/3}Si_{2/3})$ content increases to 25, the FCC phase disappears and only BCC phase was detected. Al has a much larger atomic radius than the transition metal elements of the alloys and it is known as a potent BCC phase stabilizer in steels. Figure 3 shows the hysteresis loops of the $(Fe_{0.3}Co_{0.5}Ni_{0.2})_{100-X}(Al_{1/3}Si_{2/3})_X$ series alloys at room temperature. The saturation magnetization of these alloys decreases gradually from 168.06 Am^2/kg to 116.48 Am^2/kg with increasing $(Al_{1/3}Si_{2/3})$ contents.

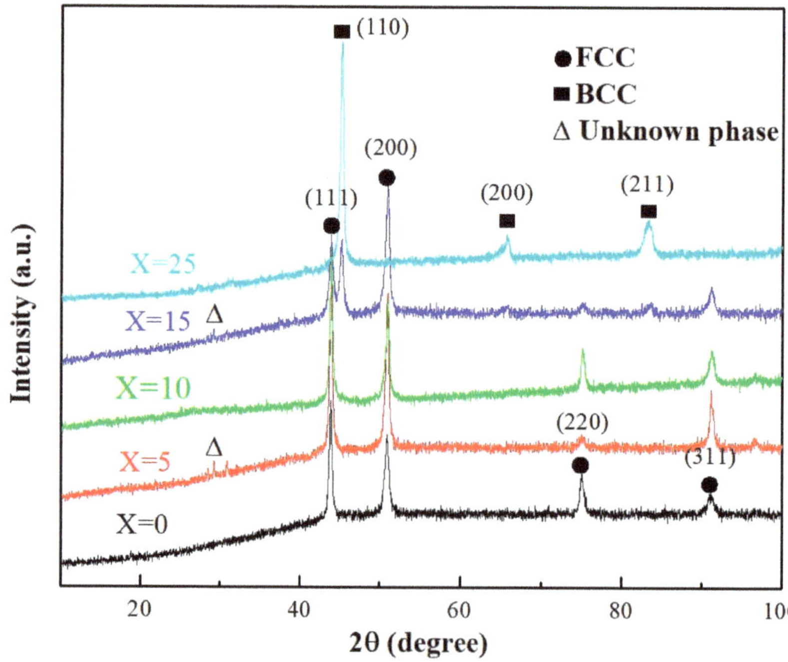

Figure 2. X-ray diffraction (XRD) patterns of $(Fe_{0.3}Co_{0.5}Ni_{0.2})_{100-X}(Al_{1/3}Si_{2/3})_X$ high-entropy alloys (HEAs) prepared using arc melting.

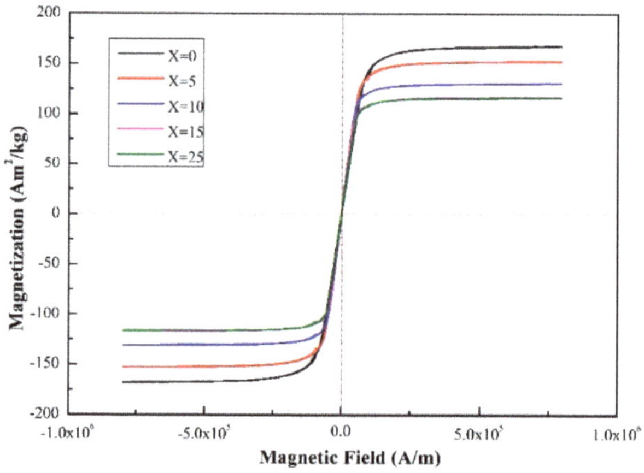

Figure 3. The hysteresis loops of $(Fe_{0.3}Co_{0.5}Ni_{0.2})_{100-X}(Al_{1/3}Si_{2/3})_X$ HEAs at room temperature.

The variation of the coercivity (Hc) and saturation magnetization (Ms) are shown in Figure 4. When X $\leq$ 10, the alloys maintain the FCC structure, and the coercivity increases with increasing Al and Si contents, increasing from 888 A/m to 1194 A/m. With higher Al and Si contents, the alloys become BCC dominant, and the Hc drops to 921 A/m when X = 15 but increases to 990 A/m when X = 25. Within a specific crystal structure, increasing Al and Si contents increase the atomic size difference and hence lattice distortion and coercivity.

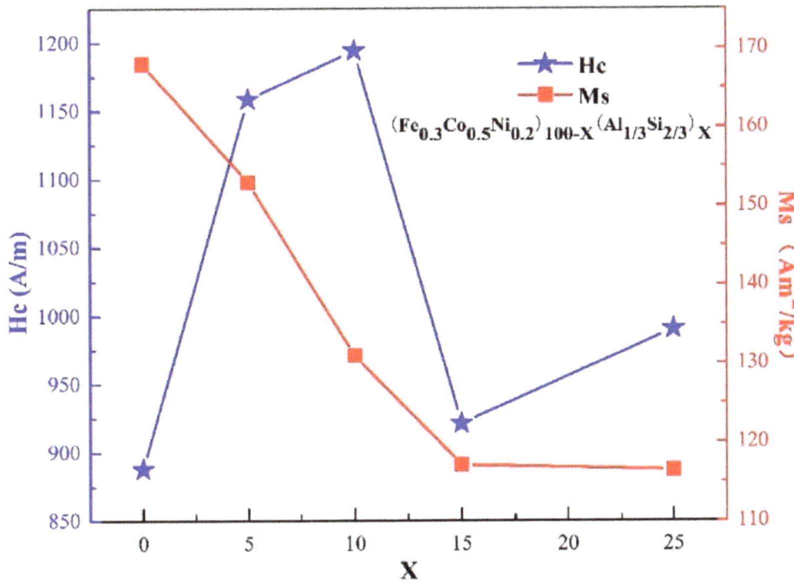

Figure 4. The variation of coercivity and saturation magnetization of $(Fe_{0.3}Co_{0.5}Ni_{0.2})_{100-X}(Al_{1/3}Si_{2/3})_X$ HEAs as a function of Al and Si contents.

The compression stress–strain behavior of $(Fe_{0.3}Co_{0.5}Ni_{0.2})_{100-X}(Al_{1/3}Si_{2/3})_X$ alloys is shown in Figure 5, and the corresponding yield stress and plastic strain are summarized in Table 1. In general, the yield stress of the alloys increases with increasing the $(Al_{1/3}Si_{2/3})$ content. When X = 5, the alloy maintains good plasticity with certain work hardening ability, and no fracture occurred during compression test. When X $\geq$ 15, the yield stress increases rapidly with a large reduction in plasticity.

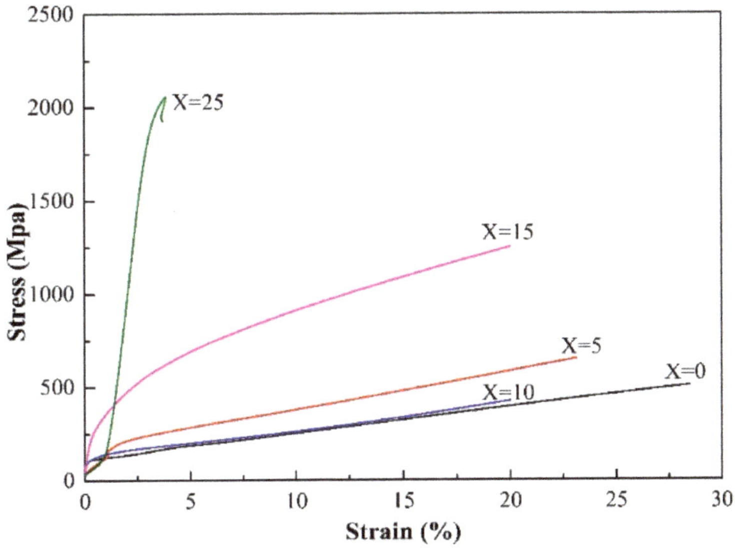

Figure 5. The compressive stress-strain curves of $(Fe_{0.3}Co_{0.5}Ni_{0.2})_{100-X}(Al_{1/3}Si_{2/3})_X$ HEAs at room temperature.

Table 1. The magnetic properties and compression behavior of $(Fe_{0.3}Co_{0.5}Ni_{0.2})_{100-X}(Al_{1/3}Si_{2/3})_X$ high entropy alloys at room temperature.

Alloys	Hc (A/m)	Ms (Am^2/kg)	$\sigma_{0.2}$ (MPa)	σ_{max} (MPa)	ε_p (%)
X = 0	887.93	168.06	96	-	>20
X = 5	1158.37	153.45	122	-	>20
X = 10	1194.29	130.97	116	-	>20
X = 15	921.25	117.10	243	-	>20
X = 25	990.48	116.48	1445	2050	2.71

The variation of mechanical properties is attributed to the different phase structures and microstructures. Dissolution of Al and Si in the FCC lattice of transition metals of Fe, Co, and Ni causes severe lattice distortion, contributing to pronounced solid solution strengthening. However, excessive addition of Al and Si promotes the formation of BCC phase, and the strength of the alloy increases rapidly at the cost of significantly reduced plasticity. Similar trend on the effect of Al contents on the phase stability and mechanical properties of high entropy alloys is reported by Tang et al [48].

For industrial applications, it is necessary for HEAs to possess not only good soft magnetic properties, but also acceptable plastic deformation ability. The newly-designed $(Fe_{0.3}Co_{0.5}Ni_{0.2})_{95}(Al_{1/3}Si_{2/3})_5$ alloy in the present study has shown good potential for industrial application as high-performance soft magents. The saturation magnetization of the alloy reaches 153.45 Am^2/kg and the resistivity is about 60.3 μΩ·cm. Besides, it also shows satisfactory ductility. A comprehensive characterization of the aloy is presented in the following sections.

Figure 6 shows the macroscopic appearance of $(Fe_{0.3}Co_{0.5}Ni_{0.2})_{95}(Al_{1/3}Si_{2/3})_5$ alloy by vacuum magnetic suspension melting and the sheet after cold-rolled. Figure 7 shows the X-ray diffraction pattern of the alloy in the as-cast, cold-rolled, water quenching, and furnace cooling states after heat treatment. A simple FCC structure can be clearly identified in all states while a small peak at around 30 degrees appears in the as-cast and furnace-cooled states.

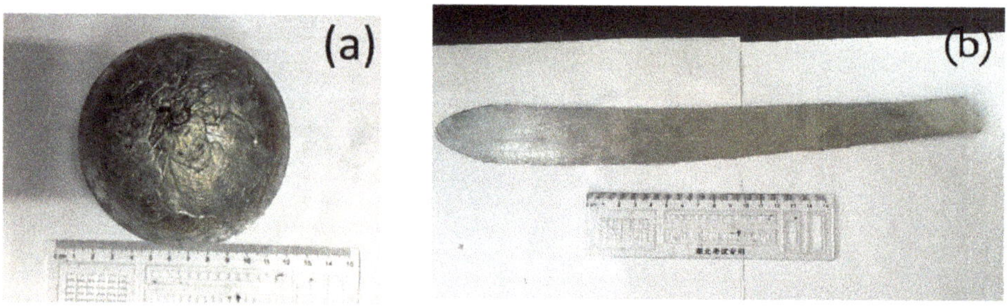

Figure 6. The macroscopic appearance: (**a**) $(Fe_{0.3}Co_{0.5}Ni_{0.2})_{95}(Al_{1/3}Si_{2/3})_5$ alloy ingot prepared by maglev melting, (**b**) cold-rolled sheet.

Figure 8 presents the EBSD images of the $(Fe_{0.3}Co_{0.5}Ni_{0.2})_{95}(Al_{1/3}Si_{2/3})_5$ alloy in different conditions. The as-cast (Figure 8a) alloy shows a coarse column grain microstructure with the column width of ~1000 μm. After cold-rolling (83% thickness reduction), the grains are significantly elongated (Figure 8b). Annealing at 1000 °C caused complete recrystallization and the resulting grain sizes were significantly smaller than in the as-cast and cold-rolled states (Figure 8c,d). Annealing also introduced the formation of a large number of annealing twins but there was no preferred grain orientation observed. Under the condition of furnace cooling, the average grain size is ~30 μm.

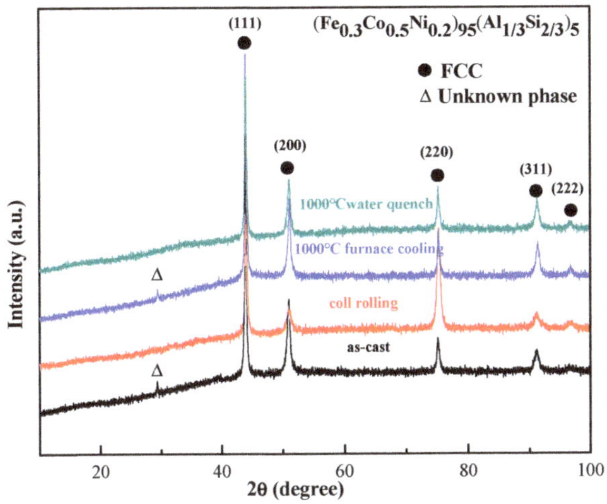

Figure 7. XRD spectra of the $(Fe_{0.3}Co_{0.5}Ni_{0.2})_{95}(Al_{1/3}Si_{2/3})_5$ alloy produced using maglev melting at different states: As-cast, cold rolling, and annealing at 1000 °C followed by furnace cooling or water quenching.

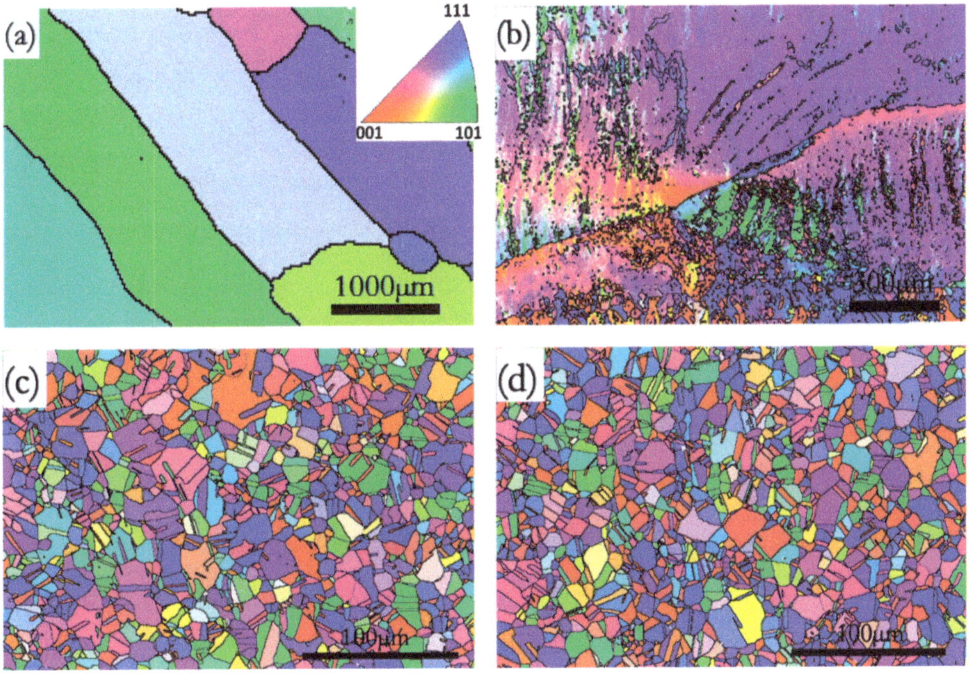

Figure 8. Electron backscatter diffraction (EBSD) images of the $(Fe_{0.3}Co_{0.5}Ni_{0.2})_{95}(Al_{1/3}Si_{2/3})_5$ alloy: (**a**) as-cast, (**b**) cold rolling, (**c**) water quenching, and (**d**) furnace cooling after annealing at 1000 °C.

Figure 9 shows the annular samples of the $(Fe_{0.3}Co_{0.5}Ni_{0.2})_{95}(Al_{1/3}Si_{2/3})_5$ alloy for magnetic induction-coercivity (B-H) measurement. Figure 10 shows the B-H and magnetization-coercivity (M-H) curves of the alloy in the states of as-cast (magnetic suspension melting), cold-rolled, water cooling, and furnace cooling after heat treatment, respectively. The soft magnetic properties of the alloy in the cold rolled variant are the worst. After the annealing treatment, the coercivity decreased from 510 A/m to 130 A/m due to internal stress release, and the saturation magnetic induction (Bs) improved from 1.14 T to 1.58 T. However, the saturation magnetization is almost constant under cold rolling and heat treatment conditions, which is about 1.49 T. Note that the coercivity of the alloy under vacuum magnetic suspension melting (96 A/m) is significantly lower than that of arc melting (>1000 A/m). The vacuum magnetic suspension melting method is cleaner than arc melting, and causes less oxide inclusions or other impurity in the ingot, and hence introduces much fewer magnetic domain pinning points in the alloy, leading to lower coercivity.

Figure 11 shows the hysteresis loss curve of the $(Fe_{0.3}Co_{0.5}Ni_{0.2})_{95}(Al_{1/3}Si_{2/3})_5$ alloy in different states. Under the power frequency conditions (50 Hz 1 T), the hysteresis loss (Ps) of the alloy in the states of as-cast, cold-rolled, water cooling, and furnace cooling after annealing is 3.13 W/kg, 8.91 W/kg, 4.26 W/kg, and 4.43 W/kg, respectively. In the application of soft magnetic materials, hysteresis loss energy will convert into heat, which will increase the temperature of the equipment and reduce working efficiency. Therefore, the hysteresis loss should be kept as small as possible.

Figure 12 presents the room temperature tensile stress-strain curves of the $(Fe_{0.3}Co_{0.5}Ni_{0.2})_{95}(Al_{1/3}Si_{2/3})_5$ alloy in the as-cast, cold-rolled, and annealed states. The specific values are listed in the Table 2. It shows that the as-cast alloy has the lowest strength, but with satisfactory ductility up to 25% elongation at room temperature. For the cold-rolled alloy, it has the lowest elongation of 1.3% but the highest yield and fracture stresses due to the work hardening.

Table 2. The magnetic properties and tensile behavior of $(Fe_{0.3}Co_{0.5}Ni_{0.2})_{95}(Al_{1/3}Si_{2/3})_5$ in different states.

$(Fe_{0.3}Co_{0.5}Ni_{0.2})_{95}(Al_{1/3}Si_{2/3})_5$	Hc (A/m)	Ms (T)	$\sigma_{0.2}$ (MPa)	σ_{max} (MPa)	ε_p (%)
Magnetic levitation melting	96	1.49	176.1	402.4	26.8
Cold rolling	510	1.48	779.6	917.7	1.3
Water quenching	130	1.40	261.5	558.3	38.3
Furnace cooling	165	1.47	235.5	572.4	33.2

Figure 9. The annular samples that were machined for magnetic induction-coercivity (B-H) measurements: Inner diameter is 18 mm, outer diameter is 20 mm, and thickness is 0.3 mm.

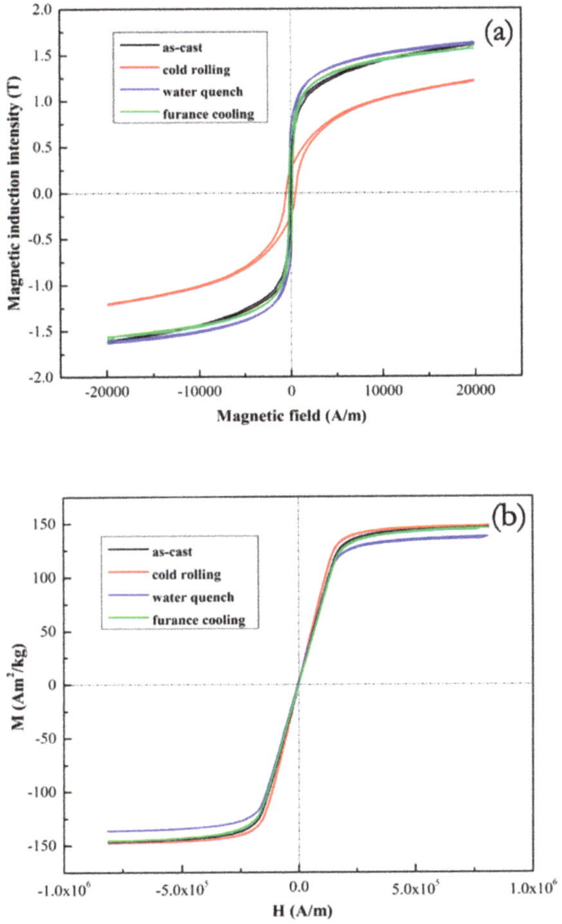

Figure 10. (**a**) magnetic induction-coercivity (B-H) and (**b**) magnetization-coercivity (M-H) (sample size: $1 \times 1 \times 4$ mm^3) curves of $(Fe_{0.3}Co_{0.5}Ni_{0.2})_{95}(Al_{1/3}Si_{2/3})_5$ alloys in the as-cast, cold rolling, and annealed states.

Magnetic materials have been developed for decades. As is well known, silicon steel is most widely used as soft magnetic material at relatively low price. However, the complex production process and large energy consumption restrict its application range [43]. New materials, such as amorphous and nanocrystalline soft magnetic materials, also have excellent soft magnetic properties, but their applications are limited by their poor mechanical properties and thermal stability. In contrast, the high entropy soft magnetic material such as $(Fe_{0.3}Co_{0.5}Ni_{0.2})_{95}(Al_{1/3}Si_{2/3})_5$ as identified in the present study can make up for the deficiency of silicon steel and amorphous soft magnetic material. It has simple preparation process and stable crystal structure, and it also has balanced magentic and mechanical properties. It can be seen from Figure 13, the $(Fe_{0.3}Co_{0.5}Ni_{0.2})_{95}(Al_{1/3}Si_{2/3})_5$ alloy reported in this study (marked by the star) presents better magnetic properties than other soft magnetic high entropy alloys.

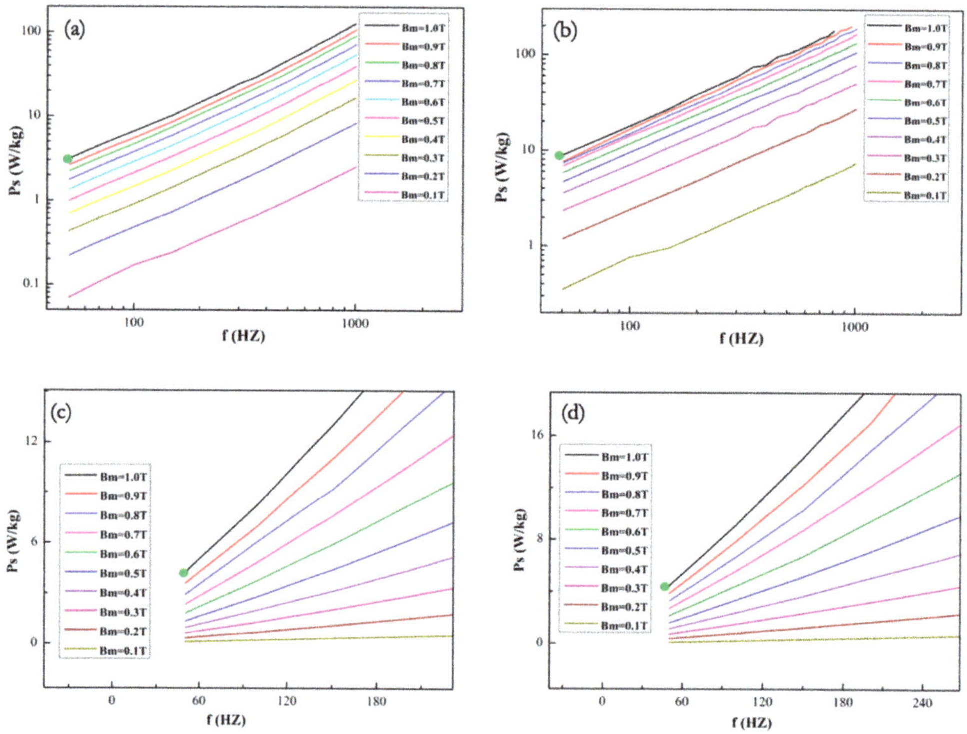

Figure 11. Power loss curves of the $(Fe_{0.3}Co_{0.5}Ni_{0.2})_{95}(Al_{1/3}Si_{2/3})_5$ alloy: (**a**) as-cast, (**b**) cold rolled, (**c**) water quenching, and (**d**) furnace cooling after annealing at 1000 °C.

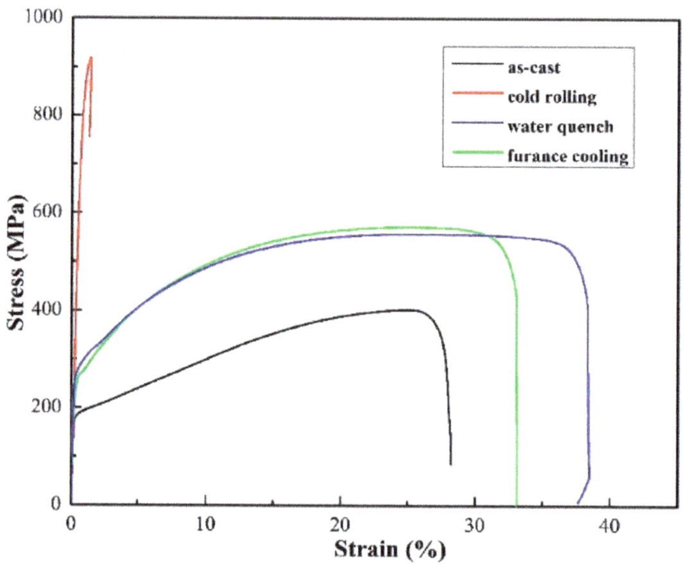

Figure 12. Tensile stress-strain curves of the $(Fe_{0.3}Co_{0.5}Ni_{0.2})_{95}(Al_{1/3}Si_{2/3})_5$ alloys in the as-cast, cold rolling and annealed states.

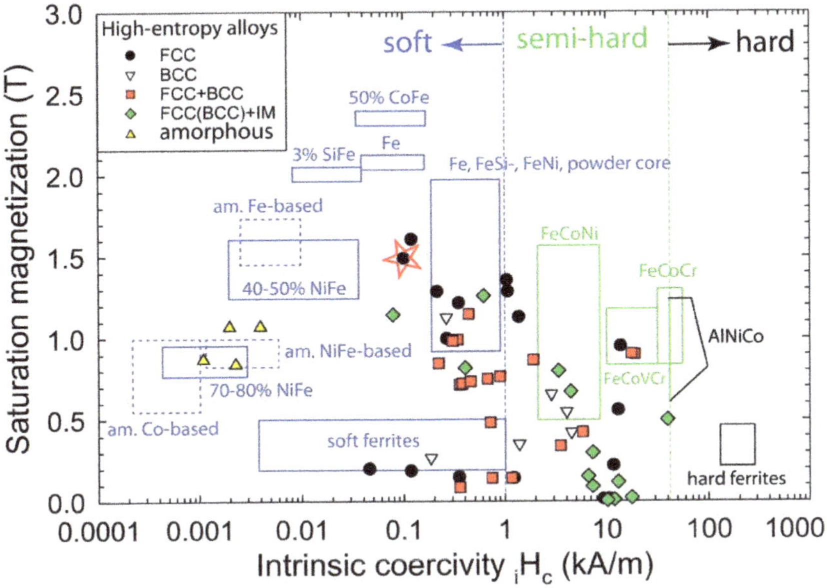

Figure 13. The saturation magnetization and intrinsic coercivity of the $((Fe_{0.3}Co_{0.5}Ni_{0.2})_{95}(Al_{1/3}Si_{2/3})_5$ alloy (marked by the red star) compared with other HEAs, amorphous metals and commercialized soft magnets compiled by Gao et al. Reproduced from [40] with permission of Cambridge University Press.

The relationship between magnetostrictive coefficient and coercivity of alloys for amorphous metals and HEAs reported in the literature is presented in Figure 14. This relationship indicates that identifying alloys with minimum saturation magnetostriction coefficient while optimizing the processing to lower coercivity is very important in developing soft magnetic materials.

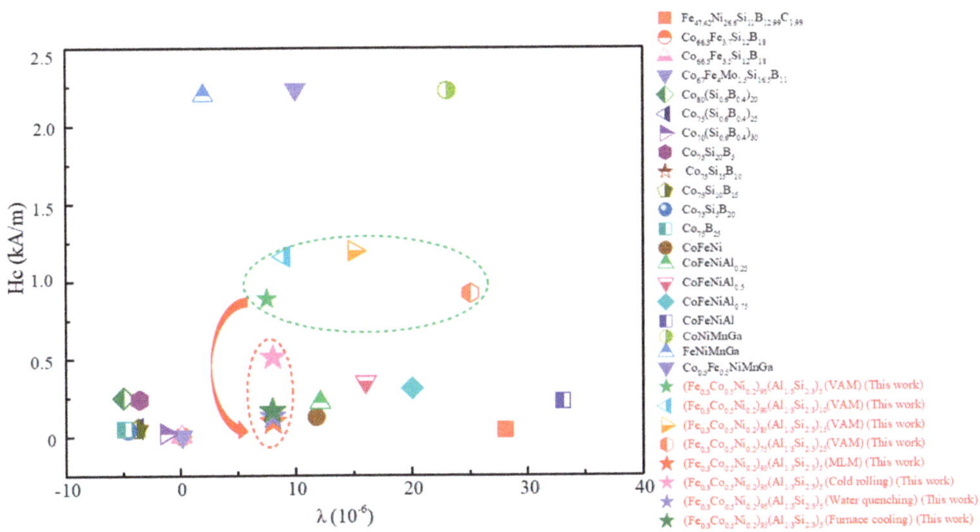

Figure 14. The relationship between saturation magnetostrictive coefficient and coercivity of amorphous alloys and HEAs (data from [46,47,49–52]).

4. Conclusions

Aiming to develop high-performance soft magnets, the magnetic and mechanical properties of a series of $(Fe_{0.3}Co_{0.5}Ni_{0.2})_{100-X}(Al_{1/3}Si_{2/3})_X$ high-entropy alloys (X = 0, 5, 10, 15, and 25) were systematically investigated in this study. The main conclusions are drawn as follows:

(1) The $Fe_{0.3}Co_{0.5}Ni_{0.2}$ base alloy was chosen since it has very small saturation magnetostriction coefficient.

(2) The alloys in the as-cast state maintain an FCC solid solution structure when $X \leq 10$. More addition of $Al_{1/3}Si_{2/3}$ causes formation of BCC phase. The overall trend is that increasing $Al_{1/3}Si_{2/3}$ contents increases the strength of the alloy at the cost of reduced ductility.

(3) The $(Fe_{0.3}Co_{0.5}Ni_{0.2})_{95}(Al_{1/3}Si_{2/3})_5$ alloy was prepared via maglev melting and then underwent cold-rolling and subsequent heat treatment. The alloy shows high saturation magnetic induction and saturation magnetization. The lower coercive force (about 100 A/m) is found in as-cast and annealed states. The hysteresis loss at low power frequency is about 3-4.5 W/kg.

(4) The $(Fe_{0.3}Co_{0.5}Ni_{0.2})_{95}(Al_{1/3}Si_{2/3})_5$ alloy in the as-cast and after annealing exhibits considerable deformation ability (ductility up to 25% and 38% elongation, respectively) at room temperature.

(5) The $(Fe_{0.3}Co_{0.5}Ni_{0.2})_{95}(Al_{1/3}Si_{2/3})_5$ alloy shows high saturation magnetic induction intensity, low coercivity, and excellent ductility. Hence, it holds good potential as next-generation high-performance soft magnetic materials for wide ranges of energy-saving applications.

Author Contributions: Conceptualization, Y.Z. and T.S.; methodology, M.Z.; formal analysis, T.Z. and D.L.; investigation, M.Z., K.Z., and B.S.; data curation, M.C.G.; writing—original draft preparation, Y.Z., M.Z., and D.L.; writing—review and editing, Y.Z., D.L., and T.S.; supervision, Y.Z.; project administration, Y.Z.

Funding: Y.Z. would like to thank the financial supports from the National Science Foundation of China (NSFC, Granted Nos. 51671020).

Acknowledgments: The technical support from the X.K. Xi of Institute of Physics, Chinese Academy of Sciences. M.C.G. acknowledges the support of the US Department of Energy's Fossil Energy Crosscutting Technology Research Program under the RSS contract 89243318CFE000003.

Conflicts of Interest: The authors declare no conflict of interest.

References

1. Yeh, J.-W. Recent progress in high-entropy alloys. *Ann. Chim. Sci. Mat.* **2006**, *31*, 633–648. [CrossRef]
2. Zhang, Y.; Zuo, T.T.; Tang, Z.; Gao, M.C.; Dahmen, K.A.; Liaw, P.K.; Lu, Z.P. Microstructures and properties of high-entropy alloys. *Prog. Mater Sci.* **2014**, *61*, 1–93. [CrossRef]
3. Miracle, D.B.; Senkov, O.N. A critical review of high entropy alloys and related concepts. *Acta Mater* **2017**, *122*, 448–511. [CrossRef]
4. Lu, Z.P.; Wang, H.; Chen, M.W.; Baker, I.; Yeh, J.W.; Liu, C.T.; Nieh, T.G. An assessment on the future development of high-entropy alloys: summary from a recent workshop. *Intermetallics* **2015**, *66*, 67–76. [CrossRef]
5. Lu, Y.; Dong, Y.; Guo, S.; Jiang, L.; Kang, H.; Wang, T.; Wen, B.; Wang, Z.; Jie, J.; Cao, Z.; et al. A promising new class of high-temperature alloys: eutectic high-entropy alloys. *Sci. Rep.* **2014**, *4*, 6200. [CrossRef]
6. Lilensten, L.; Couzinié, J.P.; Perrière, L.; Bourgon, J.; Emery, N.; Guillot, I. New structure in refractory high-entropy alloys. *Mater. Lett.* **2014**, *132*, 123–125. [CrossRef]
7. Tsai, M.-H.; Yeh, J.-W. High-entropy alloys: a critical review. *Mate. Res. Lett.* **2014**, *2*, 107–123. [CrossRef]
8. Tasan, C.C.; Diehl, M.; Yan, D.; Bechtold, M.; Roters, F.; Schemmann, L.; Zheng, C.; Peranio, N.; Ponge, D.; Koyama, M.; Tsuzaki, K.; Raabe, D. An overview of dual-phase steels: advances in microstructure-oriented processing and micromechanically guided design. *Annu. Rev. Mater. Res.* **2015**, *45*, 391–431. [CrossRef]
9. Lyu, Z.; Lee, C.; Wang, S.-Y.; Fan, X.; Yeh, J.-W.; Liaw, P.K. Effects of constituent elements and fabrication methods on mechanical behavior of high-entropy alloys: A review. *Metall. Mater. Trans. A* **2019**, *50*, 1–28. [CrossRef]
10. Gao, M.C.; Yeh, J.-W.; Liaw, P.K.; Zhang, Y. *High-Entropy Alloys*; Springer: Berlin, Germany, 2016.

11. Luo, H.; Li, Z.; Lu, W.; Ponge, D.; Raabe, D. Hydrogen embrittlement of an interstitial equimolar high-entropy alloy. *Corros. Sci.* **2018**, *136*, 403–408. [CrossRef]
12. Shun, T.-T.; Du, Y.-C. Microstructure and tensile behaviors of fcc Al0.3cocrfeni high entropy alloy. *J. Alloys Compd.* **2009**, *479*, 157–160. [CrossRef]
13. Basu, S.; Li, Z.; Pradeep, K.G.; Raabe, D. Strain rate sensitivity of a trip-assisted dual-phase high-entropy alloy. *Front. Mater. Struct. Mater.* **2018**, *5*, 1–10. [CrossRef]
14. Gu, J.; Song, M. Annealing-induced abnormal hardening in a cold rolled crmnfeconi high entropy alloy. *Scr. Mater.* **2019**, *162*, 345–349. [CrossRef]
15. Li, D.; Li, C.; Feng, T.; Zhang, Y.; Sha, G.; Lewandowski, J.J.; Liaw, P.K.; Zhang, Y. High-entropy Al 0.3 cocrfeni alloy fibers with high tensile strength and ductility at ambient and cryogenic temperatures. *Acta Mater.* **2017**, *123*, 285–294. [CrossRef]
16. Feuerbacher, M.; Lienig, T.; Thomas, C. A single-phase bcc high-entropy alloy in the refractory Zr-Nb-Ti-V-Hf system. *Scr. Mater.* **2018**, *152*, 40–43. [CrossRef]
17. Yang, T.; Zhao, Y.L.; Tong, Y.; Jiao, Z.B.; Wei, J.; Cai, J.X.; Han, X.D.; Chen, D.; Hu, A.; Kai, J.J. Multicomponent intermetallic nanoparticles and superb mechanical behaviors of complex alloys. *Science* **2018**, *362*, 933–937. [CrossRef]
18. Zhao, Y.J.; Qiao, J.W.; Ma, S.G.; Gao, M.C.; Yang, H.J.; Chen, M.W.; Zhang, Y. A hexagonal close-packed high-entropy alloy: the effect of entropy. *Mater. Des.* **2016**, *96*, 10–15. [CrossRef]
19. Lu, L.; Chen, X.; Huang, X.; Lu, K. Revealing the maximum strength in nanotwinned copper. *Science* **2009**, *323*, 607–610. [CrossRef]
20. Gao, M.C.; Zhang, B.; Guo, S.M.; Qiao, J.W.; Hawk, J.A. High-entropy alloys in hexagonal close-packed structure. *Metall. Mater. Trans. A* **2016**, *47*, 3322–3332. [CrossRef]
21. Nagase, T.; Takeuchi, A.; Amiya, K.; Egami, T. Solid state amorphization of metastable $Al_{0.5}$TiZrPdCuNi high entropy alloy investigated by high voltage electron microscopy. *Mater. Chem. Phys.* **2018**, *210*, 291–300. [CrossRef]
22. Huang, H.; Wu, Y.; He, J.; Wang, H.; Liu, X.; An, K.; Wu, W.; Lu, Z. Phase-transformation ductilization of brittle high-entropy alloys via metastability engineering. *Adv. Mater.* **2017**, *29*, 1701678. [CrossRef]
23. Li, Z.; Tasan, C.C.; Pradeep, K.G.; Raabe, D. A trip-assisted dual-phase high-entropy alloy: grain size and phase fraction effects on deformation behavior. *Acta Mater.* **2017**, *131*, 323–335. [CrossRef]
24. Senkov, O.N.; Wilks, G.B.; Miracle, D.B.; Chuang, C.P.; Liaw, P.K. Refractory high-entropy alloys. *Intermetallics* **2010**, *18*, 1758–1765. [CrossRef]
25. Miracle, D.; Miller, J.; Senkov, O.; Woodward, C.; Uchic, M.; Tiley, J. Exploration and development of high entropy alloys for structural applications. *Entropy* **2014**, *16*, 494–525. [CrossRef]
26. Tong, Y.; Chen, D.; Han, B.; Wang, J.; Feng, R.; Yang, T.; Zhao, C.; Zhao, Y.L.; Guo, W.; Shimizu, Y. Outstanding tensile properties of a precipitation-strengthened FeCoNiCrTi0.2 high-entropy alloy at room and cryogenic temperatures. *Acta Mater.* **2019**, *165*, 228–240. [CrossRef]
27. Sun, S.J.; Tian, Y.Z.; Lin, H.R.; Yang, H.J.; Dong, X.G.; Wang, Y.H.; Zhang, Z.F. Achieving high ductility in the 1.7 GPa grade cocrfemnni high-entropy alloy at 77 K. *Mater. Sci. Eng. A* **2019**, *740–741*, 336–341. [CrossRef]
28. Liu, K.; Komarasamy, M.; Gwalani, B.; Shukla, S.; Mishra, R.S. Fatigue behavior of ultrafine grained triplex $Al_{0.3}$CoCrFeNi high entropy alloy. *Scr. Mater.* **2019**, *158*, 116–120. [CrossRef]
29. Gwalani, B.; Gorsse, S.; Choudhuri, D.; Zheng, Y.; Mishra, R.S.; Banerjee, R. Tensile yield strength of a single bulk $Al_{0.3}$CoCrFeNi high entropy alloy can be tuned from 160 MPa to 1800 MPa. *Scr. Mater.* **2019**, *162*, 18–23. [CrossRef]
30. Couzinié, J.P.; Dirras, G. Body-centered cubic high-entropy alloys: from processing to underlying deformation mechanisms. *Mater. Charact.* **2019**, *147*, 533–544. [CrossRef]
31. Li, D.; Gao, M.C.; Hawk, J.A.; Zhang, Y. Annealing effect for the $Al_{0.3}$CoCrFeNi high-entropy alloy fibers. *J. Alloys Compd.* **2019**, *778*, 23–29. [CrossRef]
32. Zuo, T.; Gao, M.C.; Ouyang, L.; Yang, X.; Cheng, Y.; Feng, R.; Chen, S.; Liaw, P.K.; Hawk, J.A.; Zhang, Y. Tailoring magnetic behavior of $CoFeMnNi_x$ (X = Al, Cr, Ga, and Sn) high entropy alloys by metal doping. *Acta Mater.* **2017**, *130*, 10–18. [CrossRef]
33. Zuo, T.; Yang, X.; Liaw, P.K.; Zhang, Y. Influence of bridgman solidification on microstructures and magnetic behaviors of a non-equiatomic FeCoNiAlSi high-entropy alloy. *Intermetallics* **2015**, *67*, 171–176. [CrossRef]

34. Zhang, Y.; Zuo, T.; Cheng, Y.; Liaw, P.K. High-entropy alloys with high saturation magnetization, electrical resistivity, and malleability. *Sci. Rep.* **2013**, *3*, 1455. [CrossRef] [PubMed]
35. Lucas, M.S.; Mauger, L.; Munoz, J.A.; Xiao, Y.; Sheets, A.O.; Semiatin, S.L.; Horwath, J.; Turgut, Z. Magnetic and vibrational properties of high-entropy alloys. *J. Appl. Phys.* **2011**, *109*, 07E307. [CrossRef]
36. Singh, S.; Wanderka, N.; Kiefer, K.; Siemensmeyer, K.; Banhart, J. Effect of decomposition of the Cr–Fe–Co rich phase of alcocrcufeni high entropy alloy on magnetic properties. *Ultramicroscopy* **2011**, *111*, 619–622. [CrossRef] [PubMed]
37. Hu, F.-X.; Shen, B.-G.; Sun, J.-R. Magnetic entropy change in $Ni_{51.5}Mn_{22.7}Ga_{25.8}$ alloy. *Appl. Phys. Lett.* **2000**, *76*, 3460–3462. [CrossRef]
38. Lucas, M.S.; Belyea, D.; Bauer, C.; Bryant, N.; Michel, E.; Turgut, Z.; Leontsev, S.O.; Horwath, J.; Semiatin, S.L.; Mchenry, M.E. Thermomagnetic analysis of fecocr X Ni alloys: magnetic entropy of high-entropy alloys. *J. Appl. Phys.* **2013**, *113*, 17A923. [CrossRef]
39. Madni, A.M.; Ross, A. Exploring concept trade-offs. In *Using Tradeoff Analyses to Identify Value and Risk*; Wiley: New York, NY, USA, 2016.
40. Gao, M.C.; Miracle, D.B.; Maurice, D.; Yan, X.; Zhang, Y.; Hawk, J.A. High-Entropy functional materials. *J. Mater. Res.* **2018**, *33*, 3138–3155. [CrossRef]
41. Gutfleisch, O.; Willard, M.A.; Brück, E.; Chen, C.H.; Sankar, S.G.; Liu, J.P. Magnetic materials and devices for the 21st century: stronger, lighter, and more energy efficient. *Adv. Mater.* **2011**, *23*, 821–842. [CrossRef]
42. Garibaldi, M.; Ashcroft, I.; Simonelli, M.; Hague, R. Metallurgy of high-silicon steel parts produced using selective laser melting. *Acta Mater.* **2016**, *110*, 207–216. [CrossRef]
43. Krings, A.; Boglietti, A.; Cavagnino, A.; Sprague, S. Soft magnetic material status and trends in electric machines. *IEEE Trans. Ind. Electron.* **2017**, *64*, 2405–2414. [CrossRef]
44. Kruzic, J.J. Bulk metallic glasses as structural materials: A review. *Adv. Eng. Mater.* **2016**, *18*, 1308–1331. [CrossRef]
45. Da Silveira, G.; Slack, N. Exploring the trade-off concept. *Inter. J. Oper. Prod. Manage.* **2001**, *21*, 949–964. [CrossRef]
46. Zuo, T.T.; Li, R.B.; Ren, X.J.; Zhang, Y. Effects of Al and Si addition on the structure and properties of CoFeNi equal atomic ratio alloy. *J. Magn. Magn. Mater.* **2014**, *371*, 60–68. [CrossRef]
47. Zhang, M. Soft magnetic properties of $(Fe_{0.3}Co_{0.5}Ni_{0.2})$ 100-X $(Al_{1/3}Si_{2/3})$ X high-entropy alloy. Master's Thesis, University of Science and Technology Beijing, Beijing, China, 1 January 2019.
48. Tang, Z.; Gao, M.C.; Diao, H.; Yang, T.; Liu, J.; Zuo, T.; Zhang, Y.; Lu, Z.; Cheng, Y.; Zhang, Y. Aluminum alloying effects on lattice types, microstructures, and mechanical behavior of high-entropy alloys systems. *JOM* **2013**, *65*, 1848–1858. [CrossRef]
49. Vasquez, M.; Ascasibar, E.; Hernando, A.; Nielsen, O.V. Co-Si-B and Fe-Co-B amorphous alloys: induced anisotropy and various magnetic properties. *J. Magn. Magn. Mater.* **1987**, *66*, 37–44. [CrossRef]
50. Hernando, B.; Sanchez, M.L.; Prida, V.M.; Tejedor, M.; Vázquez, M. Magnetoimpedance effect in amorphous and nanocrystalline ribbons. *J. Appl. Phys.* **2001**, *90*, 4783–4790. [CrossRef]
51. Caltun, O.; Rao, G.S.N.; Rao, K.H.; Parvatheeswara Rao, B.; Dumitru, I.; Kim, C.-O.; Kim, C. The influence of Mn doping level on magnetostriction coefficient of cobalt ferrite. *J. Magn. Magn. Mater.* **2007**, *316*, e618–e620. [CrossRef]
52. Zhukov, A.; Churyukanova, M.; Kaloshkin, S.; Semenkova, V.; Gudoshnikov, S.; Ipatov, M.; Talaat, A.; Blanco, J.M.; Zhukova, V. Effect of annealing on magnetic properties and magnetostriction coefficient of Fe-Ni-based amorphous microwires. *J. Alloys Compd.* **2015**, *651*, 718–723. [CrossRef]

metals

MDPI

Article

Microstructures and Tribological Properties of TiC Reinforced FeCoNiCuAl High-Entropy Alloy at Normal and Elevated Temperature

Tie Zhu [1], Hong Wu [1,2,3,*], Rui Zhou [1], Ningyi Zhang [4], Yong Yin [1], Luxin Liang [1], Yong Liu [1], Jia Li [5], Quan Shan [6], Qingxiang Li [2] and Weidong Huang [3]

1 State Key Laboratory of Powder Metallurgy, Central South University, Changsha 410083, China
2 Shenzhen Zhong Jin Ling Nan Nonfemet Co., Ltd., Shenzhen 518040, China
3 School of Materials Science and Engineering, Northwestern Polytechnical University, Xi'an 710072, China
4 College of Mechanical and Vehicle Engineering, Hunan University, Changsha 410082, China
5 State Key Laboratory of Advanced Design and Manufacturing for Vehicle Body, Hunan University, Changsha 410082, China
6 School of Materials Science and Engineering, Kunming University of Science and Technology, Kunming 650093, China
* Correspondence: hwucsu@csu.edu.cn; Tel.: +86-731-8887-7669

Received: 16 January 2020; Accepted: 4 March 2020; Published: 18 March 2020

Abstract: Recent studies have suggested that high-entropy alloys (HEAs) possess high fracture toughness, good wear resistance, and excellent high-temperature mechanical properties. In order to further improve their properties, a batch of TiC-reinforced FeCoNiCuAl HEA composites were fabricated by mechanical alloying and spark plasma sintering. X-ray diffractometry analysis of the TiC-reinforced HEA composites, combined with scanning electron microscopy imaging, indicated that TiC particles were uniformly distributed in the face-centered cubic and body-centered cubic phases. The room temperature hardness of the FeCoNiCuAl HEA was increased from 467 to 768 HV with the addition of TiC, owing to precipitation strengthening and fine grain strengthening effects. As the TiC content increased, the friction coefficient of the FeCoNiCuAl HEA first increased and then decreased at room temperature, due to the transition of the wear mechanism from adhesive to abrasive behavior. At higher temperature, the friction coefficient of the FeCoNiCuAl HEA monotonously reduced, corresponding well with the transition from adhesive wear to oxidative wear.

Keywords: high-entropy alloy; TiC; tribological properties; wear mechanism

1. Introduction

In recent years, a novel class of solid-solution metallic alloys, named high-entropy alloys (HEAs) or multi-primary element alloys (MPEAs), has attracted widespread attention [1]. They usually contain five components ranging from 5% to 35% (atomic percentage, at%). Compared to traditional alloys, the HEAs exhibit various excellent properties in their strength, wear resistance, thermally-stable microstructure, oxidation, and corrosion resistance [2–4], indicating large potential applications at high temperature in engineering fields.

The microstructures and properties of the FeCoNiCuAl HEA have been widely studied over the past decade [5–7]. The addition of the aluminum (Al) element can not only increase lattice distortion and elastic energy, but also promote the formation of the body-centered cubic (BCC) phase. Therefore, it is helpful to increase the strength and hardness of HEAs. Simultaneously, the density of HEAs is significantly reduced by adding lightweight aluminum. Moreover, the addition of the copper (Cu) element can stabilize the face-centered cubic (FCC) phase and improve the hardness, ductility, and wear resistance of HEAs [8,9]. Recently, it was reported that the addition of Cu can reduce the wear rate of CoCrFeNiCux HEAs at both room temperature and elevated temperatures, and wear resistance at elevated temperatures is promoted more significantly than that at room temperature due to their self-lubricating mechanism [10].

However, the major challenge for practical applications of the FeCoNiCuAl HEA is its insufficient strength. Recent studies have indicated that the precipitation of reinforcing phases in various materials, such as HEAs and ceramics, can further increase the yield strength of the alloy while maintaining relatively high ductility [11–13]. For instance, the precipitation of SiC nanoparticles in FeCoCrNiMn enhanced the compressive yield strength from 1180 to 1480 MPa at room temperature [14]. The precipitation of WC in the FeCoCrNi HEA leads to an improved hardness up to 768HV, which corresponds well with the transition of the wear mechanism [13]. The alloy may also be strengthened by dispersion strengthening, which can significantly improve the strength and hardness of the alloy, and reduce the plasticity and toughness slightly [15]. TiC is considered a good candidate as the strengthening phase due to its high melting point, high hardness, low density, good metal matrix wettability, good chemical stability, and excellent wear resistance [16]. The optimized yield strength of the (FeCrNiCo)$Al_{0.75}Cu_{0.25}$ HEA reinforced by TiC particles can reach 1637 MPa, which is increased by 90.6% compared to the (FeCrNiCo)$Al_{0.75}Cu_{0.25}$ HEA [17]. However, the precipitation of ceramic nanoparticles in the metal matrix tends to be unevenly distributed, which is probably originated from the instability of the interfaces between the ceramic particles and the metal matrix [18–20]. In order to make the second phase uniformly distributed in the matrix, in situ methods are widely used to fabricate composites with a nanosized ceramic strengthening phase [21,22]. In general, a composite material is fabricated by an in-situ reaction of additives and carbon-based abrasives via the method of powder metallurgy [23]. Particularly, spark plasma sintering (SPS), a powder metallurgy process, provides fast processing of the feedstock, the suppression of grain coarsening and the production of specimens with a low porosity [24,25]. The combination of the ultra-fine grain and nanosized reinforcing particles can enhance the strength of the composite material through the Hall–Petch effect and the strengthening of dislocations [26,27]. At the same time, the in-situ method can make the interfaces between the ceramic and the metal substrate cleaner, showing stronger bonding forces between materials that strengthen the mechanical properties of the composite material [28].

As demonstrated above, the precipitation strengthening effects in HEAs have been extensively investigated [13,14,29–33]. However, the high-temperature tribological properties of ceramic-reinforced HEAs have rarely been reported. In the present study, FeCoNiCuAl HEA composites reinforced by in situ-generated TiC ceramic particles were fabricated by mechanical alloying (MA) followed by SPS. The microstructures, phase compositions, and hardness of the FeCoNiCuAl-TiC composites were characterized, respectively. The effect of strengthening phases on the high-temperature tribological properties of the HEA was studied, and the related mechanism was discussed.

2. Experimental

The pre-alloyed powder of the FeCoNiCuAl HEA was prepared using gas atomization in a high-purity argon atmosphere to avoid oxidation. The particle size of the HEA powder was measured by a laser particle size analyzer (MASTERSIZER3000, Malvern, Worcestershire, UK). The average particle size of the HEA powder was approximately 9.2 μm. The equiatomic-ratio FeCoNiCuAl HEA was mixed with varying compositions (0 wt%, 10 wt%, 20 wt%, 30 wt%) of Ti powder and graphite nanoparticles (around 100 nm) in a high-energy planetary ball milling machine (MITR YXQM-4L, MITR, Changsha, China) for 5 h. The milling speed was 250 rpm, and the ball-to-powder weight ratio was 10:1. Ethanol was used as a processing control agent to avoid cold welding. The as-milled powders were then consolidated by spark plasma sintering (SPS; FCT D25/3) in a 40 mm graphite die at a sintering temperature of 1000 °C under 30 MPa pressure for 10 min. Graphite foils with applied BN coating were placed between the powders and the die walls to avoid possible contamination. The as-SPSed bulk specimens were polished using standard metallographic silicon carbide papers supplemented with colloidal silica powders (Struers OPS).

The phase composition of the HEA-TiC composite was analyzed using a D/max 2500pc X-ray diffractometer (XRD) (Advance D8, Bruker, Karlsruhe, Germany) with Cu–Ka radiation (k = 0.154 nm). The microstructure observation was conducted by a scanning electron microscope (SEM) equipped with energy dispersive spectrometer (EDS) and electron backscatter diffraction (EBSD) (FEI Nova Nano230). The hardness of the specimens was measured by a micro-Vickers hardness tester at a load of 5 kg for 15 s (HVS-5). The micro-hardness of each phase was determined by an Ultra Nanoindenter (UNHTL +MCT, CSM, Switzerland) under the load of 30 mN for 10 s. The wear tests were performed using a HT-1000 ball-on-disk tribometer (HT-1000, Lanzhou ZhongKeKaiHua Technology Development Co. Ltd, Lanzhou, China) under constant applied normal load of 10 N at a line speed of 0.5 m/s for 30 min and at the testing temperatures of 20 and 600 °C, respectively. Commercial Si_3N_4 balls were used as the counterface. The diameter of the Si_3N_4 balls was 6 mm. The volumetric loss of the alloy was measured using a contact surface profilometer (Keyence VHX-5000). The volume wear rate was calculated by the formula $W = V/SL$, where W is the volume wear rate, V is the volumetric wear loss, S is the sliding distance, and L is the normal load. To ensure reproducibility, the wear tests were performed at least three times at both room and high temperature. The mean value of the wear rates was calculated. The worn surfaces of the specimens were examined by SEM.

3. Results

3.1. Microstructures

Figure 1a shows XRD patterns of the HEA composite powders after ball milling. The HEA powders consisted mainly of FCC and BCC phases. As for the as-milled HEA-TiC powders, diffraction peaks of Ti and graphite were distinctly observed. The peak intensity of Ti and graphite increased with the increase in Ti and graphite contents. The XRD patterns of SPSed specimens are shown in Figure 1b. Likewise, the SPSed HEA still comprised FCC and BCC phases. In the HEA-TiC composites, the peaks of Ti and graphite disappeared instead of TiC peaks, which indicate the formation of TiC during the sintering process. Table 1 presents the mixed enthalpy between carbon and several metal elements. Considering the lowest value of mixed enthalpy between Ti and C, the formation of TiC is reasonable and consistent with that reported by Yim [17]. According to the XRD data, the equilibrium lattic e constants of the FCC and BCC phases in the HEA were calculated as 3.521 and 2.881 Å, respectively, while the lattice constants of FCC and BCC phases in the composites were 3.53 and 2.884 Å, respectively. This discrepancy is probably due to the solid solution effect of the C atoms in the matrix [21,22].

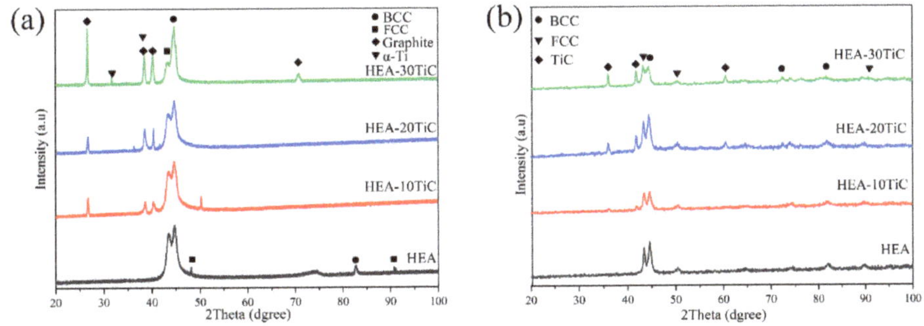

Figure 1. XRD patterns of (**a**) HEA-TiC composite as-milled powders and (**b**) as-SPSed HEA-TiC composites.

Table 1. The mixing enthalpy of different atomic pairs [34].

ΔH_{mix}, kJ/mol	Fe	Co	Ni	Cu	Al	Ti
C	−50	−42	−39	−33	−36	−109

During the MA process, oxidation inevitably occurs and oxides are commonly included in the alloys [35]. Figure 2 presents the microstructures of the sintered HEAs. In addition to the alloy matrix, there are two types of impurities, i.e., white particles attached on the surface (Figure 2a) and black spots dispersed in the matrix (Figure 2b). The white particles are presumed to be silica powders in oxide polishing solution (OPS) owing to their similar size (0.04 µm), while the black spots correspond to oxides [17,36]. The impurities are generally introduced by the milling medium and/or the process controlling agent (PCA) during the MA process [37,38].

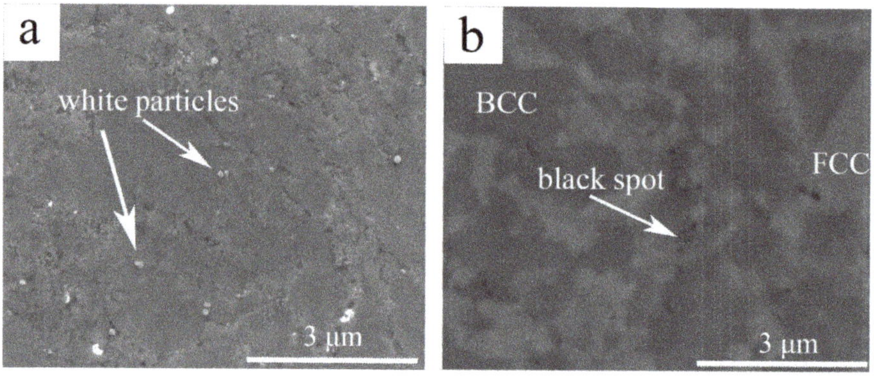

Figure 2. (**a**) Scanning electron (SE) and (**b**) backscattered electron (BSE) images of the microstructures of the SPSed HEA.

As shown in Figure 2b, the HEA consists mainly of a gray phase and a white phase, corresponding to the FCC and BCC phases, respectively, which is in accord with the results reported by Beyramali Kivy et al. [8]. It was reported that Ni and Al show high concentrations in the BCC phase, while Cu was abundant in the FCC phase [39]. The EDS analysis results of the bright and dark areas in Figure 2b are listed in Table 2. According to this result, the crystal structure of the bright area was FCC phase and the gray area was BCC phase. The microstructures of HEA-TiC composites with different TiC contents are presented in Figure 3. A number of black dots with a diameter of 0.3 µm were uniformly distributed in the FCC and BCC phases. The composition of the black phase is exhibited in Figure 4, indicating the formation of TiC particles in the HEA. The distribution of TiC particles in the HEA was uneven; it is easier to precipitate in the FCC phase due to the FCC structure of TiC [29]. In the HEA-TiC composites, with increasing Ti and graphite contents, the content of TiC increased.

Table 2. Chemical compositions of the FCC phase, BCC phase, and TiC phase.

Chemical Composition	Fe	Co	Ni	Cu	Al	Ti	C
FCC	26.7	21.4	17.0	27.0	7.3	-	-
BCC	23.2	24.0	25.0	7.8	19.7	-	-
TiC	5.1	5.3	5.3	2.2	3.3	53.3	25.6

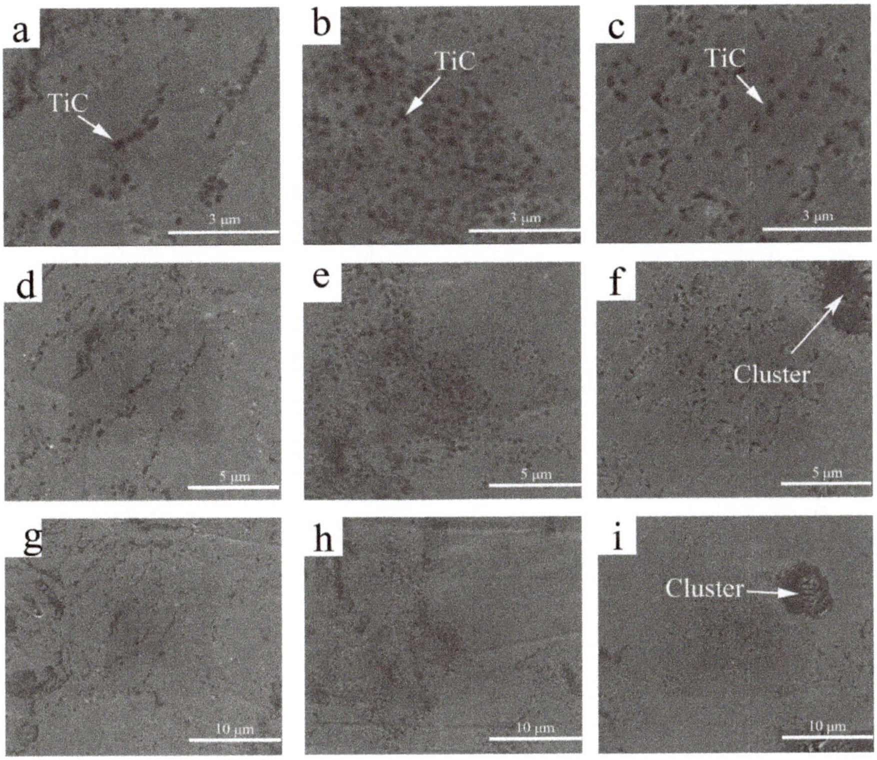

Figure 3. SE images of the microstructures of the SPSed (**a**,**d**,**g**) HEA-10TiC, (**b**,**e**,**h**) HEA-20TiC, and (**c**,**f**,**i**) HEA-30TiC composites.

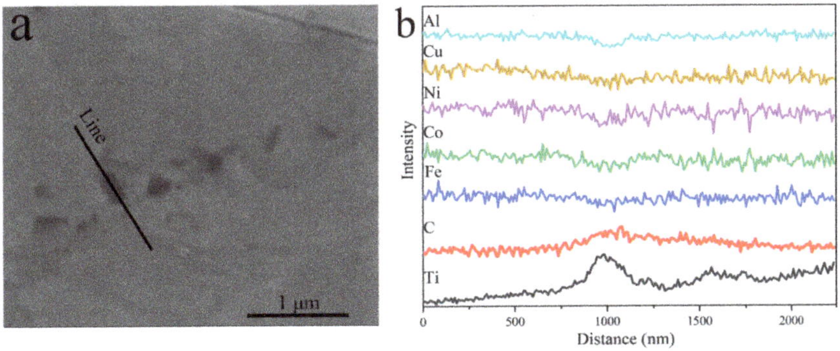

Figure 4. (**a**) SEM image showing the position of the line scan and (**b**) compositional profiles of the Ti- and C-rich carbide in the HEA-10TiC composite.

Figure 5 exhibits the EBSD inverse pole figure (IPF) maps of the HEA and the 10 wt% HEA-TiC composite. TiC particles were observed in the XRD patterns (Figure 1) and SEM images (Figure 3). However, they were not detected in EBSD maps due to the limited particle size of TiC. It is apparent that the average grain size of the 10 wt% HEA-TiC composite is smaller than that of the HEA. The addition of TiC can suppress the growth of grains during sintering, leading to the refinement of the grains in the HEA matrix. Figure 5c,d present the phase compositions of the HEA and the HEA-10TiC composite, respectively. The FeCoNiCuAl HEA is mainly composed of the BCC phase and a small amount of the FCC phase. With the addition of Ti and graphite, the content of FCC increases from 11.9% in the HEA to 19.2% in the HEA-10TiC composite.

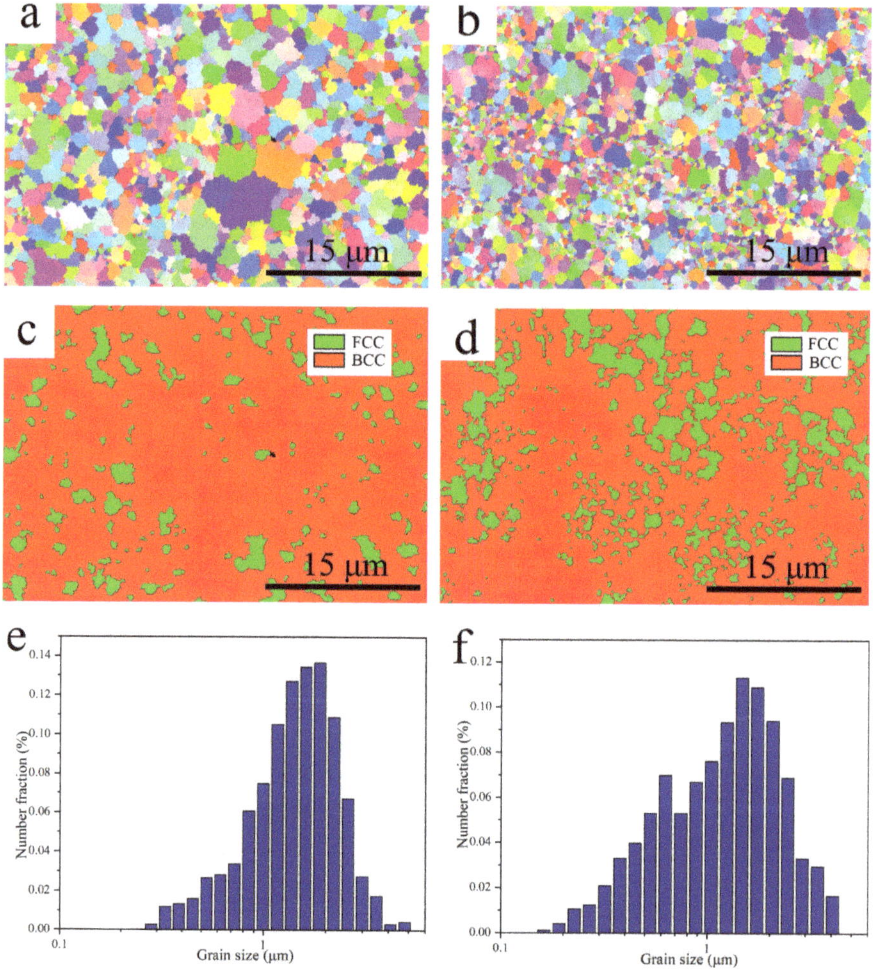

Figure 5. Inverse pole figure (IPF) maps of the SPSed (**a**) HEA and (**b**) HEA-10TiC, phase composition of the SPSed (**c**) HEA and (**d**) HEA-10TiC, and grain size of the SPSed (**e**) HEA and (**f**) HEA-10TiC.

3.2. Hardness

The hardness of the HEA-TiC composites is shown in Figure 6a. As the content of TiC increases, the hardness of the composites remarkably increases from 467 to 768 HV. The micro-hardness of the matrix phase, TiC particles, and the interface between TiC particles is illustrated in Figure 6b. The result indicated that the agglomerated TiC exhibited the highest micro-hardness (1489 HV) among all phases. The micro-hardness of the interface was about 916 HV, and the matrix phase showed the lowest micro-hardness of only 570 HV. The high hardness of the TiC particles may contribute to the improvement of the wear resistance of the HEA-TiC composites. Figure 6c presents typical loading–unloading curves for indentations with a 50 mN load. The TiC particle showed a higher elastic modulus than that of the matrix phase.

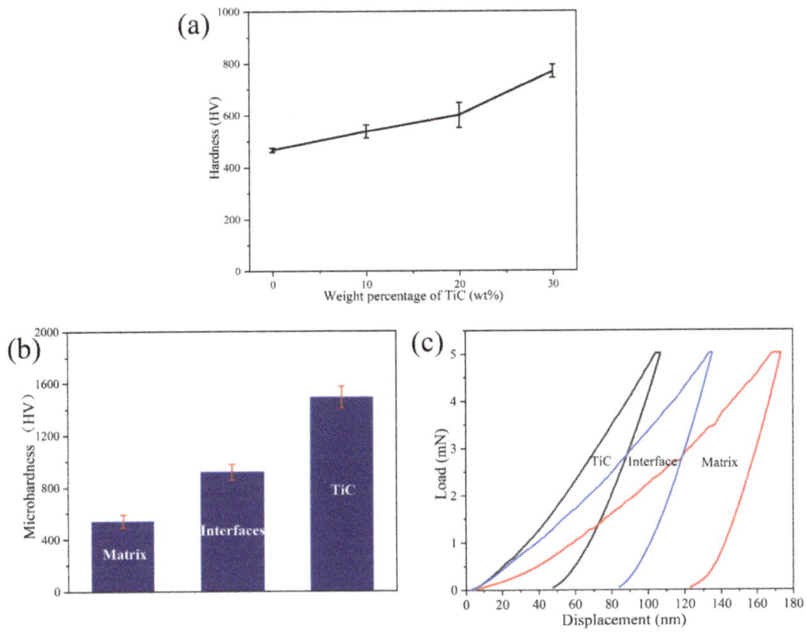

Figure 6. (**a**) Hardness of HEA-TiC composites; (**b**) micro-hardness of different regions in the HEA-10TiC composites; (**c**) load–displacement curves of different regions in the HEA-10TiC composites.

3.3. Friction and Wear Properties, Worn Surface, and Debris

The friction and wear properties of the HEA-TiC composites are shown in Figure 7. The friction coefficients of all specimens sharply increased at the beginning of the wear tests and reached a steady state corresponding to the breaking-in stage. Subsequently, all of the friction coefficient curves remained relatively stable and exhibited periodic waves, which may be owing to periodic accumulation, elimination of debris and the periodic localized fracture of the surface layer [40]. The increase in the friction coefficient was caused by the accumulation of abrasive debris on the wear surface, and the separation of the wear debris from the surface or the filling of the wear cracks caused the reduction of the friction coefficient. As shown in Figure 7b, the average friction coefficient increased at first and then decreased with further addition of Ti and graphite to 30 wt%. The volumetric wear rate curves of the composites at room temperature are shown in Figure 7c. With the increase in the Ti and graphite contents, the hardness of the composite increased and the wear rate decreased.

Figure 8 shows the worn surface morphologies as well as the debris of HEA-TiC composites. The typical morphologies of both adhesive wear and abrasive wear were observed in the HEAs. When the stress arising from the grinding ball on the worn surface was higher than the yield strength of the HEAs, the deformation process was triggered. Under alternating stress, debris was torn off from the surface and deformed as flakes, as shown in Figure 8e. The EDS results of the worn surface are summarized in Table 3. The oxygen contents of the worn surface increased slightly with the addition of Ti and graphite, indicating that no considerable oxidation occurred. Figure 8c, d show comparatively rough surface generated by the micro-cutting and micro-plowing of abrasive particles. The furrow-like wear scar along the sliding direction in HEA-TiC composites implied an abrasive wear dominated process. The grinding chips were either particle-like or flake-shaped. With the increase of TiC content, the amount of flaky wear debris decreased, while the particle-like debris increased (Figure 8e–h).

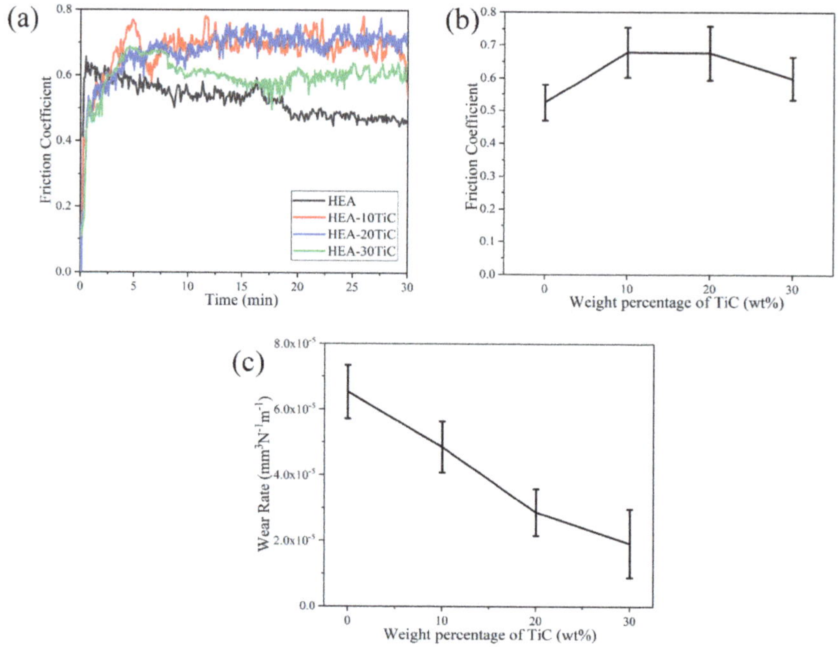

Figure 7. (**a**) Friction coefficient vs. sliding time. (**b**) Friction coefficient vs. TiC content. (**c**) Wear rate of HEA-TiC composites at room temperature.

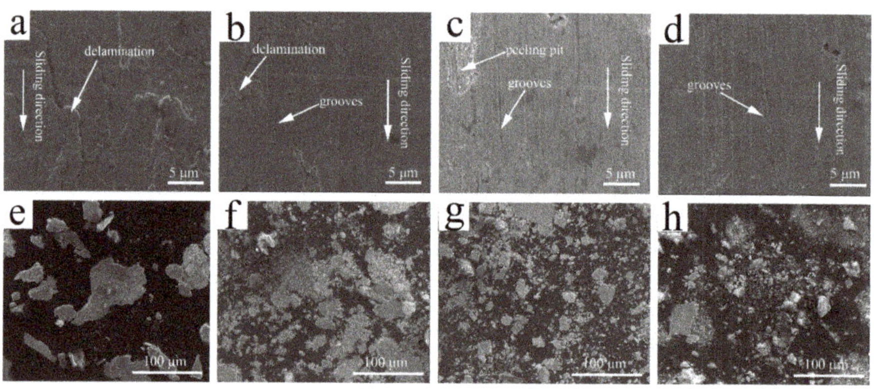

Figure 8. SEM images of the worn surface of (**a**) HEA, (**b**) HEA-10TiC, (**c**) HEA-20TiC, and (**d**) HEA-30TiC composites and corresponding debris of (**e**) HEA, (**f**) HEA-10TiC, (**g**) HEA-20TiC, and (**h**) HEA-30TiC composites after wear at room temperature.

Table 3. Chemical compositions of the surface wear-tested at room temperature, in atomic percentage.

Chemical Composition	Fe	Co	Ni	Cu	Al	Ti	C	O
TiC00	21.3	22.0	21.2	11.0	19.6	-	-	4.9
TiC10	16.1	13.5	12.3	17.9	12.5	8.5	12.7	6.6
TiC20	15.5	15.3	14.8	8.8	13.7	14.4	10.7	7.0
TiC30	10.7	10.3	9.8	9.2	8.6	24.0	15.9	11.4

The friction and wear properties of the HEA-TiC composites at high temperature are shown in Figure 9. As shown in Figure 9a, the friction coefficient of the HEA was unstable and exhibited drastic fluctuation, which may be related to the aggregation of large debris. In contrast, the friction coefficients of the HEA-TiC composites remained comparatively stable after the run-in period. With the addition of Ti and graphite, the average friction coefficient dropped considerably (Figure 9b). As plotted in Figure 9c, the wear rate of the HEA was significantly higher than that of the HEAs-TiC composites. However, it clearly presented only small wear scatters among the HEA-TiC composites, which is mainly ascribed to the transition of the wear mechanism. This will be further clarified in the Discussion section. Recent work by Moravcikova-Gouvea et al. [41] reported that a HEA produced by MA and SPS exhibited better wear resistance than traditional AISI 52,100 and Inconel 713 alloys, and suggested that powder metallurgy provides a good approach to fabricating HEAs with fine-grained microstructures and enhanced wear resistance. Similarly, in the present study, the wear rates of the HEA-TiC composites have the same order of magnitude (10^{-5} mm^3/Nm) at room temperature, while the high-temperature wear rates of the TiC-containing specimens are dramatically lower than that of the TiC-free specimen, and even other HEAs, by at least one order of magnitude [42,43].

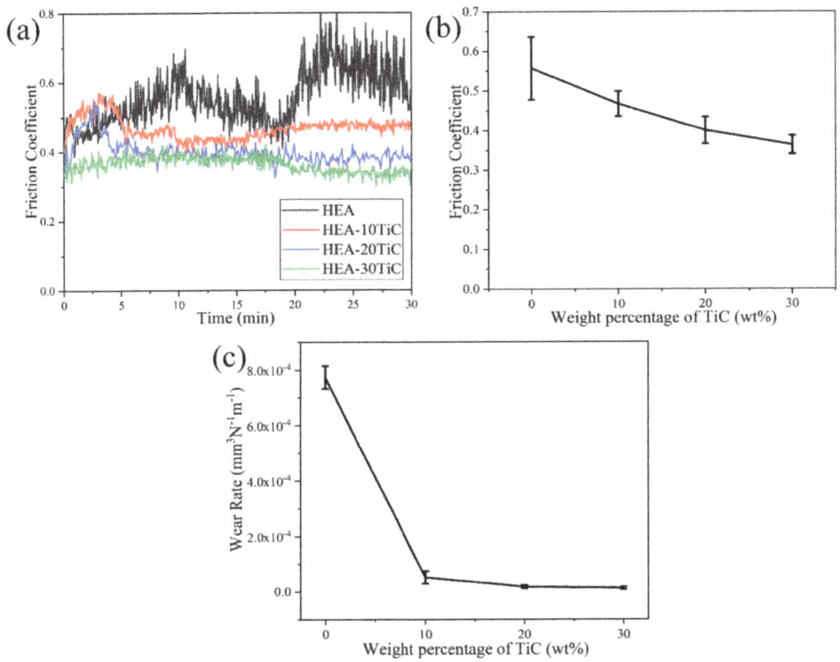

Figure 9. (**a**) Friction coefficient vs. sliding time. (**b**) Friction coefficient vs. TiC content (**c**). Wear rate of HEA-TiC composites at high temperature.

Figure 10 exhibits the worn surface and debris of the HEA-TiC composites at high temperature (600 °C). The worn surface of the HEA was smooth (Figure 10a), and large pieces of flake-like debris were peeled off from the wear surface (Figure 10e). These indicated the occurrence of severe adhesive wear. In contrast, the high-temperature worn surfaces of the HEA-TiC composites were relatively smoother, and smearing of the wear scars occurred (Figure 10b–d). The material of interface contact points was soft, and a metal-like film on the friction surface was formed. The metal-like film can act as a lubricating agent, and result in a considerably reduced friction coefficient. The grinding chips generated by wear tests performed at high temperature were primarily particle-like (Figure 10f–h). With the increase of TiC content, the amount of wear debris decreased and the size reduced.

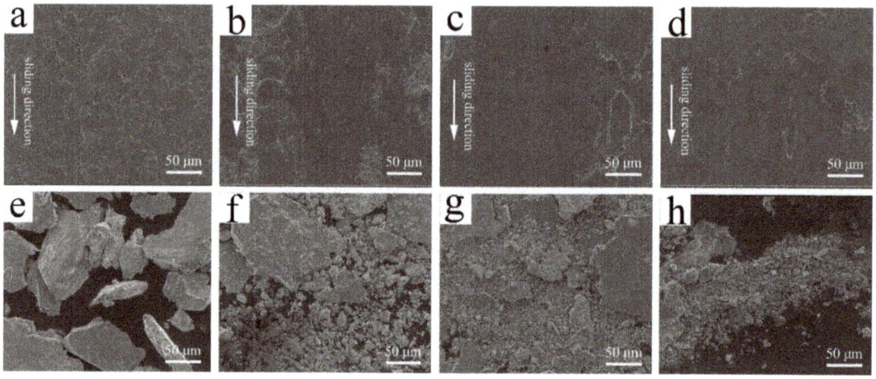

Figure 10. SEM images of the worn surface of (**a**) HEA, (**b**) HEA-10TiC, (**c**) HEA-20TiC, and (**d**) HEA-30TiC composites and corresponding debris of (**e**) HEA, (**f**) HEA-10TiC, (**g**) HEA-20TiC, and (**h**) HEA-30TiC composites after wear at high temperature.

4. Discussion

4.1. Microstructures

It is well accepted that the microstructure plays a significant role in the wear resistance of materials. In the present study, the microstructure of the FeCoNiCuAl HEA was mainly composed of the FCC and BCC phases. The EDS results indicated that Co was almost uniformly distributed in both phases compared to other elements. Ni and Al showed relatively higher concentrations in the BCC phase, while Cu and Fe were abundant in the FCC phase. These are consistent with other reports [8,9,44]. The precipitation mechanism for the ordered Al–Ni-rich matrix phase is spinodal decomposition due to the large mixing enthalpy between Al and Ni [29,45]. With the addition of Ti and graphite, the grain size reduced. As shown in Figure 3, the segregation of the TiC ceramic particles in the HEA matrix may be attributed to the dual-phase structure [29]. The volume fraction of TiC increased accompanied by the increasing incorporation of Ti and graphite. Large clusters were observed in the HEA-30TiC composites (Figure 3F,I).

4.2. Wear Behavior

During the sliding processes, the HEA-TiC composites exhibited excellent wear resistance. The reason for this may be explained as follows. First, TiC, acting as a strengthening phase, can effectively reduce the plastic deformation of the matrix during sliding, thus improving the wear resistance of the composites [46]. Second, the grain sizes of the HEA-TiC composites were lower than that of the HEA. The TiC particles can inhibit the growth of grains to obtain finer grains, thereby enhancing the strength and hardness. According to Archard's law, assuming that it is under the condition of adhesive wear and sliding spherical asperities deform fully plastically in contact, the wear resistance of the material is proportional to its hardness, and thus it will be promoted as well [47,48]. Thus, the wear resistance of the HEA-TiC composites was better than that of the HEA due to their high hardness [40]. Further, the excellent antioxidation properties in the composites are proposed to improve the wear resistance [49]. The EDS results of the HEA-TiC composites indicated that the existence of oxidized layers can effectively prevent direct metal–metal contact and impede adhesive wear. Therefore, the wear rate of the HEA-TiC composites was further reduced with the increase in Ti and graphite.

The EDS results exhibit that the contents of Cu and O are remarkably high (Table 4), which indicates the presence of an oxide layer, thus improving the wear properties [10,40]. During wear tests conducted at high temperature, the formation of an oxidation layer, acting as a lubricating film, is easier than the tests performed at room temperature. This implied that oxidative wear was dominated. Two competitive processes may coexist. The debris peels off from the composites, and oxidation layers are formed by cold welding. Meanwhile, the fragmentation of the oxidation layer continually occurs to produce new debris [50]. In the HEA, the debris are not well compacted, which makes them easy to remove and gives rise to the enhancement of wear rate. In the HEA-TiC composites, in contrast, the oxidation layers are strengthened by the TiC particles to produce glazed layers, which may efficiently prevent metal–metal contact and reduce friction between the HEA-TiC composites and the counterface. Consequently, the friction and wear properties of the HEA-TiC composites are substantially improved at high temperature [29].

Table 4. Chemical compositions of the surface wear-tested at high temperature, in atomic percentage.

Chemical Composition	Fe	Co	Ni	Cu	Al	Ti	C	O
TiC00	9.9	10.9	11.4	14.6	25.2	-	-	28.0
TiC10	2.6	3.5	4.1	15.5	5.4	3.4	4.8	60.6
TiC20	1.5	3.7	3.4	13.0	3.9	6.8	3.0	64.7
TiC30	1.0	2.2	3.0	15.8	3.8	9.8	4.1	60.6

5. Conclusions

(1) The FeCoNiCuAl-TiC composites mainly consist of FCC, BCC, and TiC phases. The TiC particles are dispersed in the matrix with a diameter of 0.3 μm.

(2) The micro-hardness of FeCoNiCuAl-TiC composites increases from 467 to 768 HV with the increase in TiC content, which is primarily attributed to the strengthening of the hard TiC phases and the refined microstructures.

(3) The main wear mechanism for the FeCoNiCuAl HEA at room temperature is adhesive wear. The wear mechanisms for the FeCoNiCuAl-TiC composites at room temperature are adhesive wear and abrasive wear. With the increase in TiC, the friction coefficient increases first and then decreases, while the wear loss continually decreases.

(4) The wear mechanism of the FeCoNiCuAl HEA at high temperature is severe adhesive wear. With the addition of TiC, both the friction coefficient and the wear rate reduce considerably. The dominant wear mechanism of HEA-TiC composites at high temperature is oxidative wear. Benefiting from the formation of oxidation layers, the HEA-TiC composite exhibits excellent wear resistance.

Author Contributions: Conceptualization, T.Z. and Y.Y.; methodology, H.W., J.L., and L.L.; data curation, Q.L. and W.H.; writing—original draft preparation, T.Z.; writing—review and editing, T.Z., Y.L., H.W., and R.Z.; supervision, N.Z. and Q.S. All authors have read and agreed to the published version of the manuscript.

Funding: This research was funded by the National Key Research and Development Program (Grant No. 2016YFB1100103), National Natural Science Foundation of China (Grant No. 51771233), Key Research and Development Program of Hunan Province (Grant No. 2016JC2003), China Postdoctoral Science Foundation (Grant No. 2018M633164), Fundamental Research Funds for the Central Universities of Central South University (Grant No. 2018ZZTS127, CX20190190, 2019ZZTS134), and Science and Technology Program of Shenzhen City (JSGG20170824162647398).

Acknowledgments: The authors are grateful to the National Key Research and Development Program.

Conflicts of Interest: The authors declare no conflict of interest.

References

1. Yeh, J.-W. Nanostructured high-entropy alloys with multiple principal elements. *Nov. Alloy Des. Concepts Outcomes* **2004**, *6*, 299–303. [CrossRef]

2. He, J.Y.; Wang, H.; Huang, H.L.; Xu, X.D.; Chen, M.W.; Wu, Y.; Liu, X.J.; Nieh, T.G.; An, K.; Lu, Z.P. A precipitation-hardened high-entropy alloy with outstanding tensile properties. *Acta Mater.* **2016**, *102*, 187–196. [CrossRef]
3. Liu, B.; Wang, J.; Liu, Y.; Fang, Q.; Wu, Y.; Chen, S.; Liu, C.T. Microstructure and mechanical properties of equimolar FeCoCrNi high entropy alloy prepared via powder extrusion. *Intermetallics* **2016**, *75*, 25–30. [CrossRef]
4. Poletti, M.G.; Fiore, G.; Gili, F.; Mangherini, D.; Battezzati, L. Development of a new high entropy alloy for wear resistance: FeCoCrNiW 0.3 and FeCoCrNiW 0.3 + 5 at.% of C. *Mater. Des.* **2017**, *115*, 247–254. [CrossRef]
5. Butler, T.M.; Weaver, M.L. Oxidation behavior of arc melted AlCoCrFeNi multi-component high-entropy alloys. *J. Alloys Compd.* **2016**, *674*, 229–244. [CrossRef]
6. Li, D.; Li, C.; Feng, T.; Zhang, Y.; Sha, G.; Lewandowski, J.J.; Liaw, P.K.; Zhang, Y. High-entropy Al 0.3 CoCrFeNi alloy fibers with high tensile strength and ductility at ambient and cryogenic temperatures. *Acta Mater.* **2017**, *123*, 285–294. [CrossRef]
7. Li, Q.H.; Yue, T.M.; Guo, Z.N.; Lin, X. Microstructure and corrosion properties of AlCoCrFeNi high entropy alloy coatings deposited on AISI 1045 steel by the electrospark process. *Metall. Mater. Trans. A* **2012**, *44*, 1767–1778. [CrossRef]
8. Beyramali Kivy, M.; Asle Zaeem, M.; Lekakh, S. Investigating phase formations in cast AlFeCoNiCu high entropy alloys by combination of computational modeling and experiments. *Mater. Des.* **2017**, *127*, 224–232. [CrossRef]
9. Zhuang, Y.X.; Xue, H.D.; Chen, Z.Y.; Hu, Z.Y.; He, J.C. Effect of annealing treatment on microstructures and mechanical properties of FeCoNiCuAl high entropy alloys. *Mater. Sci. Eng. A* **2013**, *572*, 30–35. [CrossRef]
10. Verma, A.; Tarate, P.; Abhyankar, A.C.; Mohape, M.R.; Gowtam, D.S.; Deshmukh, V.P.; Shanmugasundaram, T. High temperature wear in CoCrFeNiCux high entropy alloys: The role of Cu. *Scr. Mater.* **2019**, *161*, 28–31. [CrossRef]
11. Mao, C.; Zhou, F.; Hu, Y.; Cai, P.; Jiang, Y.; Bi, Z.; Peng, G. Tribological behavior of cBN-WC-10Co composites for dry reciprocating sliding wear. *Ceram. Int.* **2019**, *45*, 6447–6458. [CrossRef]
12. Mao, C.; Lu, J.; Zhao, Z.; Yin, L.; Hu, Y.; Bi, Z. Simulation and experiment of cutting characteristics for single cBN-WC-10Co fiber. *Precis. Eng.* **2018**, *52*, 170–182. [CrossRef]
13. Zhou, R.; Chen, G.; Liu, B.; Wang, J.; Han, L.; Liu, Y. Microstructures and wear behaviour of (FeCoCrNi) 1-x (WC) x high entropy alloy composites. *Int. J. Refract. Met. Hard Mater.* **2018**, *75*, 56–62. [CrossRef]
14. Rogal, Ł.; Kalita, D.; Tarasek, A.; Bobrowski, P.; Czerwinski, F. Effect of SiC nano-particles on microstructure and mechanical properties of the CoCrFeMnNi high entropy alloy. *J. Alloys Compd.* **2017**, *708*, 344–352. [CrossRef]
15. Dobeš, F.; Hadraba, H.; Chlup, Z.; Dlouhý, A.; Vilémová, M.; Matějíček, J. Compressive creep behavior of an oxide-dispersion-strengthened CoCrFeMnNi high-entropy alloy. *Mater. Sci. Eng. A* **2018**, *732*, 99–104. [CrossRef]
16. Jiang, Q.; Li, X.; Wang, H. Fabrication of TiC particulate reinforced magnesium matrix composites. *Scr. Mater.* **2003**, *48*, 713–717. [CrossRef]
17. Yim, D.; Sathiyamoorthi, P.; Hong, S.-J.; Kim, H.S. Fabrication and mechanical properties of TiC reinforced CoCrFeMnNi high-entropy alloy composite by water atomization and spark plasma sintering. *J. Alloys Compd.* **2018**. [CrossRef]
18. Du, X.; Gao, T.; Liu, G.; Liu, X. In situ synthesizing SiC particles and its strengthening effect on an Al–Si–Cu–Ni–Mg piston alloy. *J. Alloys Compd.* **2017**, *695*, 1–8. [CrossRef]
19. Shu, D.; Li, Z.; Zhang, K.; Yao, C.; Li, D.; Dai, Z. In situ synthesized high volume fraction WC reinforced Ni-based coating by laser cladding. *Mater. Lett.* **2017**, *195*, 178–181. [CrossRef]
20. Zhao, N.; Xu, Y.; Fu, Y. Mechanical properties of one-step in situ synthesized NbC-Fe composite coating. *Surf. Coat. Technol.* **2017**, *309*, 1105–1110. [CrossRef]
21. Cheng, J.; Liu, D.; Liang, X.; Chen, Y. Evolution of microstructure and mechanical properties of in situ synthesized TiC–TiB2/CoCrCuFeNi high entropy alloy coatings. *Surf. Coat. Technol.* **2015**, *281*, 109–116. [CrossRef]

22. Guo, N.N.; Wang, L.; Luo, L.S.; Li, X.Z.; Chen, R.R.; Su, Y.Q.; Guo, J.J.; Fu, H.Z. Microstructure and mechanical properties of in-situ MC-carbide particulates-reinforced refractory high-entropy Mo0.5NbHf0.5ZrTi matrix alloy composite. *Intermetallics* **2016**, *69*, 74–77. [CrossRef]
23. Moravcik, I.; Cizek, J.; Gavendova, P.; Sheikh, S.; Guo, S.; Dlouhy, I. Effect of heat treatment on microstructure and mechanical properties of spark plasma sintered AlCoCrFeNiTi0.5 high entropy alloy. *Mater. Lett.* **2016**, *174*, 53–56. [CrossRef]
24. Löbel, M.; Lindner, T.; Lampke, T. Enhanced wear behaviour of spark plasma sintered AlCoCrFeNiTi high-entropy alloy composites. *Materials* **2018**, *11*, 2225. [CrossRef] [PubMed]
25. Sakkaki, M.; Sadegh Moghanlou, F.; Vajdi, M.; Shahedi Asl, M.; Mohammadi, M.; Shokouhimehr, M. Numerical simulation of heat transfer during spark plasma sintering of zirconium diboride. *Ceram. Int.* **2020**, *46*, 4998–5007. [CrossRef]
26. Ko, Y.G.; Shin, D.H.; Park, K.-T.; Lee, C.S. An analysis of the strain hardening behavior of ultra-fine grain pure titanium. *Scr. Mater.* **2006**, *54*, 1785–1789. [CrossRef]
27. Liu, Y.; Wang, J.; Fang, Q.; Liu, B.; Wu, Y.; Chen, S. Preparation of superfine-grained high entropy alloy by spark plasma sintering gas atomized powder. *Intermetallics* **2016**, *68*, 16–22. [CrossRef]
28. Fan, Q.C.; Li, B.S.; Zhang, Y. The microstructure and properties of (FeCrNiCo)AlxCuy high-entropy alloys and their TiC-reinforced composites. *Mater. Sci. Eng. A* **2014**, *598*, 244–250. [CrossRef]
29. Sun, X.; Zhu, H.; Li, J.; Huang, J.; Xie, Z. Influence of aluminum content on the microstructure and properties of the in-situ TiC reinforced AlxFeCoNiCu high entropy alloy matrix composites. *Mater. Sci. Eng. A* **2019**, *743*, 540–545. [CrossRef]
30. Sun, X.; Zhu, H.; Li, J.; Huang, J.; Xie, Z. High entropy alloy FeCoNiCu matrix composites reinforced with in-situ TiC particles and graphite whiskers. *Mater. Chem. Phys.* **2018**. [CrossRef]
31. Rogal, Ł.; Kalita, D.; Litynska-Dobrzynska, L. CoCrFeMnNi high entropy alloy matrix nanocomposite with addition of Al_2O_3. *Intermetallics* **2017**, *86*, 104–109. [CrossRef]
32. An, Q.; Wang, J.; Liu, Y.; Liu, B.; Guo, W.; Fang, Q.; Nie, Y. Effects of C and Mo on microstructures and mechanical properties of dual-phase high entropy alloys. *Intermetallics* **2019**, *110*. [CrossRef]
33. Zhou, R.; Li, M.; Wu, H.; Liu, B.; Liu, Y. Effect of multi-component carbides on the mechanical behavior of a multi-element alloy. *Mater. Sci. Eng. A* **2019**, *758*, 99–102. [CrossRef]
34. Inoue, A.T.A.A. Classification of bulk metallic glasses by atomic size difference, heat of mixing and period of constituent elements and its application to characterization of the main alloying element. *Mater. Trans.* **2005**, *46*, 2817–2829. [CrossRef]
35. Peng, Y.B.; Zhang, W.; Mei, X.L.; Wang, H.J.; Zhang, M.Y.; Wang, L.; Li, X.F.; Hu, Y. Microstructures and mechanical properties of FeCoCrNi-Mo high entropy alloys prepared by spark plasma sintering and vacuum hot-pressed sintering. *Mater. Today Commun.* **2020**. [CrossRef]
36. Sathiyamoorthi, P.; Basu, J.; Kashyap, S.; Pradeep, K.G.; Kottada, R.S. Thermal stability and grain boundary strengthening in ultrafine-grained CoCrFeNi high entropy alloy composite. *Mater. Des.* **2017**, *134*, 426–433. [CrossRef]
37. Murty, B.S.; Ranganathan, S.J.M.R. Novel materials synthesis by mechanical alloying/milling. *Int. Mater. Rev.* **1998**, *43*, 101–141. [CrossRef]
38. Suryanarayana, C. Mechanical alloying and milling. *Prog. Mater. Sci.* **2001**, *46*, 1–184. [CrossRef]
39. Wen, L.H.; Kou, H.C.; Li, J.S.; Chang, H.; Xue, X.Y.; Zhou, L. Effect of aging temperature on microstructure and properties of AlCoCrCuFeNi high-entropy alloy. *Intermetallics* **2009**, *17*, 266–269. [CrossRef]
40. Wu, J.-M.; Lin, S.-J.; Yeh, J.-W.; Chen, S.-K.; Huang, Y.-S.; Chen, H.-C. Adhesive wear behavior of AlxCoCrCuFeNi high-entropy alloys as a function of aluminum content. *Wear* **2006**, *261*, 513–519. [CrossRef]
41. Moravcikova-Gouvea, L.; Moravcik, I.; Omasta, M.; Vesely, J.; Cizek, J.; Minarik, P.; Cupera, J.; Zadera, A.; Jan, V.; Dlouhy, I. High-strength Al0.2Co1.5CrFeNi1.5Ti high-entropy alloy produced by powder metallurgy and casting: A comparison of microstructures, mechanical and tribological properties. *Mater. Charact.* **2020**, *159*, 752–757. [CrossRef]
42. Yu, Y.; He, F.; Qiao, Z.; Wang, Z.; Liu, W.; Yang, J. Effects of temperature and microstructure on the triblogical properties of CoCrFeNiNbx eutectic high entropy alloys. *J. Alloys Compd.* **2019**, *775*, 1376–1385. [CrossRef]
43. Zhuang, Y.X.; Liu, W.J.; Chen, Z.Y.; Xue, H.D.; He, J.C. Effect of elemental interaction on microstructure and mechanical properties of FeCoNiCuAl alloys. *Mater. Sci. Eng. A* **2012**, *556*, 395–399. [CrossRef]

44. Zhuang, Y. Effect of Co element on microstructure and mechanical properties of FeCoXNiCuAl alloys.pdf. *Acta Metallurgica Sinica* **2012**. [CrossRef]
45. Kao, Y.-F.; Chen, T.-J.; Chen, S.-K.; Yeh, J.-W. Microstructure and mechanical property of as-cast, -homogenized, and -deformed AlxCoCrFeNi ($0 \leq x \leq 2$) high-entropy alloys. *J. Alloys Compd.* **2009**, *488*, 57–64. [CrossRef]
46. Du, B.; Wang, X.; Zou, Z. Microstructure and tribological behavior of laser in situ synthesized TiC-reinforced Fe-based composite coatings. *Tribol. Lett.* **2011**, *43*, 295–301. [CrossRef]
47. Zhou, Z.; Shan, Q.; Jiang, Y.; Li, Z.; Zhang, Z. Effect of nanoscale V2C precipitates on the three-body abrasive wear behavior of high-Mn austenitic steel. *Wear* **2019**, 436–437. [CrossRef]
48. Archard, J.F. Contact and rubbing of flat surfaces. *J. Appl. Phys.* **1953**, *24*, 981–988. [CrossRef]
49. Chuang, M.-H.; Tsai, M.-H.; Wang, W.-R.; Lin, S.-J.; Yeh, J.-W. Microstructure and wear behavior of AlxCo1.5CrFeNi1.5Tiy high-entropy alloys. *Acta Mater.* **2011**, *59*, 6308–6317. [CrossRef]
50. Stott, F.H. High-temperature sliding wear of metals. *Tribol. Int.* **2002**, *35*, 489–495. [CrossRef]

metals

Article

Production and Properties of High Entropy Carbide Based Hardmetals

Johannes Pötschke [1,*], Manisha Dahal [2], Anne Vornberger [1], Mathias Herrmann [1] and Alexander Michaelis [1,2]

1 Fraunhofer IKTS, Fraunhofer Institute for Ceramic Technologies and Systems, 01277 Dresden, Germany; anne.vornberger@ikts.fraunhofer.de (A.V.); mathias.herrmann@ikts.fraunhofer.de (M.H.); alexander.michaelis@ikts.fraunhofer.de (A.M.)
2 Institute of Materials Science (IfWW), Technische Universität Dresden, 01067 Dresden, Germany; dahal.manisa@gmail.com
* Correspondence: johannes.poetschke@ikts.fraunhofer.de; Tel.: +49-351-2553-7641

Abstract: Dense, high-entropy carbide cobalt-bonded hardmetals with two different compositions, namely (Hf-Ta-Ti-Nb-V)C-19.2 vol% Co and (Ta-Ti-Nb-V-W)C-19.2 vol% Co, were successfully manufactured by gas pressure sintering (SinterHIP) at 1400 °C and 100 bar Ar pressure. The microstructure of these hardmetals consists of a rigid skeletal carbide phase embedded in a tough Co binder phase. EDS mappings showed that the high-entropy carbide phase did not decompose and that a typical hardmetal microstructure was realized. Only in the case of the (Hf-Ta-Ti-Nb-V)C-Co hardmetal was some undissolved TaC and HfO_2, as well as some clustered vanadium titanium carbide phase, found, resulting in a split-up of the HEC phase into two very similar HEC phases. This resulted in a reduced hardness to fracture toughness ratio for this composition. Measurements of magnetic saturation polarization showed values between 57.5% and 70% of theoretical magnetic saturation polarization, indicating marginal dissolution of the carbide-forming metal elements in the binder phase. The hardness value HV10 for (Hf-Ta-Ti-Nb-V)C-19.2 vol% Co was 1203 HV10 and 1432 HV10 for (Ta-Ti-Nb-V-W)C-19.2 vol% Co.

Keywords: carbide; high-entropy carbides; binders; microstructure; mechanical properties; high-entropy hardmetals

Citation: Pötschke, J.; Dahal, M.; Vornberger, A.; Herrmann, M.; Michaelis, A. Production and Properties of High Entropy Carbide Based Hardmetals. *Metals* **2021**, *11*, 271. https://doi.org/10.3390/met11020271

Academic Editors: Peter K. Liaw and Weidong Li

Received: 18 December 2020
Accepted: 2 February 2021
Published: 5 February 2021

1. Introduction

Basic hardmetals (also known as cemented carbides) and cermets consist of hard carbide phases (WC or TiCN) mixed with a ductile binder phase (Co/Ni/Fe). Extensive studies on monocarbide hardmetal compositions such as WC-Co, WC-Ni, NbC-Co, NbC-Ni, VC-Ni, Cr_3C_2-Ni, etc., exist in the literature [1–6]. Other carbides such as TiC, TaC, and NbC are also present as a mixed carbide composition, such as (Ti, Ta, Nb, W)C, for better wear resistance, hardness, and toughness [7]. Studies on mixed carbides such as WC-Cr_3C_2-Co, WC-VC-Co, (Ta, Mo)C-Co/Ni, (Ta, W)C-Co/Ni [8–13] showed that the stability of these mixed carbides strongly depends on their chemical composition and solubility in the metal binder phase.

Furthermore, several cermet compositions such as (Ti-Mo)C-Co, (Ti, Mo, W)C-Co, (W-Ti)C-(Co/Ni), TiN-Ni, TiC-Mo-Ni, WC-(Ta, Nb)C, TiC-Mo_2C-(TaC-NbC-VC)Ni, and (Ti, W, Ta)C-Mo-Ni have been studied [14–21]. In cermets too, the addition of additional carbides to its TiC/TiCN-based main hard phase is mostly carried out only in small amounts for the improvement of wettability, the suppression of grain growth, and the enhancement of toughness [7]. Regardless of the composition, cermets mainly consist of a multi-phase core–rim structure with unalloyed lighter TiCN in the core surrounded by other heavier carbides in the rim [22].

Replacing traditional binders (Co/Ni/Fe or mixtures thereof) with high-entropy alloys (HEA) show a good wettability of WC, high toughness, prominent wear resistance, and high temperature stability. For example, by replacing Co with CoCrFeNi, CoCrFeMnNi or CoCrFeNiV binder, the plastic deformation resistance of hardmetals was enhanced outperforming traditional WC-Co hardmetals [23,24]. In cermets too, CoCrFeNiCu can lead to less interfaces by inhibiting core–rim structure and grain growth, thus enhancing wettability, hardness, and toughness [23].

Inspired by the concept of HEA, high entropy carbides (HEC) have gained attention in recent years. As opposed to having a principal metal carbide with minor additions of other carbides, HEC´s are composed of five or more equiatomic carbides. Several researchers have already studied a broad range of medium- and high-entropy carbide compositions, such as (Hf-Zr-Ta-Nb-Ti)C, (Hf-Ta-Zr-Nb)C, (Ti-Zr-Hf-V-Nb)C or (Hf-Zr-Ti-Ta-Nb)C [21,25–32]. These HEC´s can form a single-phase solid solution by increasing the solubility among elements due to enhanced mixing entropy, and suppress the formation of intermetallic phases and element segregations.

Spark plasma sintering (SPS), also known as field-assisted sintering (FAST), is the most used method for the production of bulk HEC so far [26,27,29,30,32]. Several HEC compositions have also been manufactured by hot pressing [31,33,34] and vacuum sintering [34]. Furthermore, the synthesis of HEC powders has been studied for several compositions [35,36]. These studies have shown that HEC possesses superior hardness, toughness, and wear resistance as compared to certain monocarbides. The knowledge of the production and properties of HEC has opened up new possibilities for tailoring a wide range of high-entropy carbide-based hardmetals. However, the properties of this newly emerging class of materials, e.g., hardmetal compositions such as HEC-Co, HEC-Ni and similar, remain so far unexplored [37].

In this study, two different pre-manufactured HEC compositions were crushed to HEC powders and mixed with Co, followed by subsequent sintering using a SinterHIP process. The influence of HEC compositions on the density, grain size, and basic mechanical properties of the obtained HEC-Co hardmetals was studied and is discussed within this work for the first time.

2. Materials and Methods

2.1. Preparation of HEC Powders

High-entropy carbides of two different compositions were manufactured using a conventional powder metallurgical route. Commercially available refractory carbides of HfC, TaC, TiC, NbC, VC, and WC were mixed in equiatomic proportions to produce two different high-entropy carbides: (Hf-Ta-Ti-Nb-V)C and (Ta-Ti-Nb-V-W)C. Their exact compositions are given in Table 1. The details on powder supplier, FSSS particle size, specific surface area (BET), d_{BET} particle size, as well as the oxygen content of the powders are given in Table 2.

The specific surface (S_{BET}) was measured under Krypton (ASAP 2020 MP, Micromeritics, Norcross, GA, USA), and the particle diameter d_{BET} was calculated using the formula $d_{BET} = 6000/(S_{BET} * \text{density})$. The mean particle size d_{FSSS} was measured with a Fisher Sub-sieve sizer (Model 95, Fisher Scientific, Waltham, MA, USA).

Table 1. Compositions of HEC as a precursor to manufacture HEC-Co.

HEC Samples	Compositions (wt. %)					
	HfC	TaC	NbC	TiC	VC	WC
HEC 1	31.2	31.6	17.2	9.8	10.3	-
HEC 2	-	31.3	17.0	9.7	10.2	31.8

Table 2. Characteristics of the used starting powders (* data from the suppliers' datasheets).

Powders	Supplier	Particle Size d_{FSSS} (μm)	Specific Surface Area BET (m^2/g)	d_{BET} (nm)	O_2 Content (wt. %)
HfC	Sigma-Aldrich	0.6	3.9	133	1.72
TaC	H.C. Starck	1.0	1.1	366	0.13 *
NbC	GTP	1.2	1.2	639	0.25 *
TiC	H.C. Starck	2.4	2.0	619	0.37 *
VC	H.C. Starck	1.2	3.0	350	0.80 *
WC	H.C. Starck	0.5	2.6	149	0.26 *

Refractory carbide powders were mixed in compositions given in Table 1, ball-milled in n-heptane, vacuum dried, and sieve granulated. The granulated powder mixtures were uniaxially pressed to bars and sintered by gas pressure sintering (FPW 280/600, FCT, Sonneberg, Germany). The gas pressure sintering cycle, here referred to as sinterHIP, was carried out at 1950 °C with a holding time of 240 min in vacuum, followed by 45 min of 100 bar Ar pressure. The two prepared HEC 1 and HEC 2 compositions (Table 1) were again crushed in a vibratory disc mill (RS 200, Retsch, Haan, Germany, WC-Co vial and disc) at 700 RPM, and sieved to a particle size of less than 20 μm.

2.2. Preparation of HEC-Co Hardmetals

The crushed HEC 1 and HEC 2 powders were mixed with Co (Umicore, half micron, d_{FSSS} = 0.7 μm, O_2 content = 0.69 wt. %) to a Co content of 19.2 vol% (equivalent binder content of a WC-12 wt. % Co hardmetal), and ball-milled for 48 h. This was followed by vacuum drying and sieve granulation. The granulated powder mixture was then uniaxially pressed to bending bars and sintered. The sinterHIP cycle was caried out at 1400 °C and an Ar pressure of 100 bar with a holding time of 45 min. Altogether, two different HEC-based hardmetals (HEC-Co) were produced: composition (Hf-Ta-Ti-Nb-V)C-Co is designated as HEC 1-Co and composition (Ta-Ti-Nb-V-W)C-Co as HEC 2-Co.

The density of all sintered samples was measured according to ISO 3369. For microstructural analysis, dense samples were ground and polished down to 1 μm using different diamond slurries, similar to what is described in ISO 4499-1. Images of microstructures were taken using the field emission scanning electron microscope (FESEM) ULTRA 55 (Carl Zeiss Microscopy, Oberkochen, Germany), which was also used to perform EDS analysis. ISO porosity was determined according to ISO 4499-4. The phase composition of the powder mixtures (as pressed) as well as of sintered samples was analyzed by X-ray diffraction (XRD) using a diffractometer D8 Advance (Bruker AXS, Karlsruhe, Germany), operated with Cu-Kα radiation with a LynxEye position-sensitive detector (PSD, Bruker AXS, Karlsruhe, Germany) and a nickel filter located in the primary beam. The Diffrac.EVA program and the JCPDS database (2020) were used for phase analysis. The magnetic properties coercivity Hc and magnetic saturation polarization mS were measured according to ISO 3326.

The Vickers hardness (HV10) of sintered samples was measured in agreement with ISO 3878 with a load of 10 kP and five measurements per sample. Fracture toughness (K_{1C}) was calculated by measuring the Vickers indentation crack length using the Shetty equation [38].

3. Results and Discussion

3.1. Synthesis of HEC Powders

The HEC powders were synthesized similar to bulk binderless HEC ceramics [39], but with slightly adjusted synthesis conditions. Figure 1a,b show the corresponding micrographs of the bulk HEC compositions HEC 1 ((Hf-Ta-Nb-Ti-V)C) and HEC 2 ((Ta-Nb-Ti-V-W)C) which were used for the HEC powder preparation.

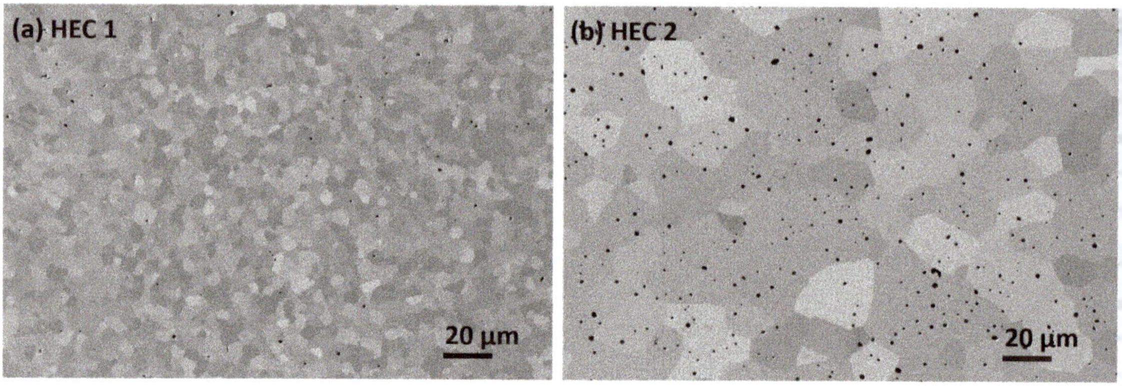

Figure 1. Micrographs of produced high-entropy carbides with sinterHIP cycle with (**a**) HEC 1 composition (Hf-Nb-Ta-Ti-V)C and (**b**) HEC 2 composition (Ta-Nb-Ti-V-W)C before crushing and sieving.

From Figure 1, differences in grain size can be observed. The grain sizes of HEC 1 ca be estimated to be in the range of approx. 2 µm to 5 µm, while for HEC 2, it ranges fro approx. 5 µm to 30 µm. After crushing the HEC powders to <20 µm, it is expected th in the case of HEC 1 some polycrystalline HEC grains remain, while in the case of HEC mostly monocrystalline HEC grains will be present.

For HEC 1, the EDS measurements showed small third phases next to the HE phase consisting of either TaC or HfO_2. In the case of HEC 2, only the HEC phase an small amounts of free carbon (black spots in Figure 1b) were detected. Neverthele semiquantitative EDS measurement and the XRD measurement (see next section) showe that the amount is too small to change the composition of the HEC significantly.

3.2. Properties of HEC-Co Hardmetals

The prepared HEC 1 and HEC 2 powders were used as a hard phase, which were the mixed with 19.2 vol% Co (equivalent to WC-12 wt. % Co). First, the samples subjecte to sinterHIP were characterized by measuring density. Theoretical density, calculate according to the rule of mixture, Archimedes density, and relative density are give in Table 3.

Table 3. Density of prepared samples HEC 1-Co and HEC 2-Co.

Hardmetal Samples	Sintering Conditions	Theoretical Density (g/cm³)	Archimedes Density (g/cm³)	Relative Density (%)
HEC 1-19.2 vol% Co	sinterHIP	9.18	9.15	99.6
HEC 2-19.2 vol% Co	1400 °C/100 bar Ar	9.64	9.74	101.0

The data in Table 3 prove that the samples are completely densified. The 101 theoretical density for HEC 2-Co is due to the inaccuracy of the calculation of theoretic density. The SEM micrographs of the polished cross section reveal that the samples ar completely densified with an ISO porosity of A02B00C00 for both compositions.

The backscattered electron micrographs of both HEC-Co hardmetals in Figure represent various phases. The dark gray represents the Co binder phase, whereas gra and light gray contrasts show the carbide grains. Both micrographs represent a skelet microstructure of cubic carbides embedded in a cobalt binder phase. Differences in gra size directly correspond to the initial grain sizes of the HEC materials used as startin materials (see Figure 1).

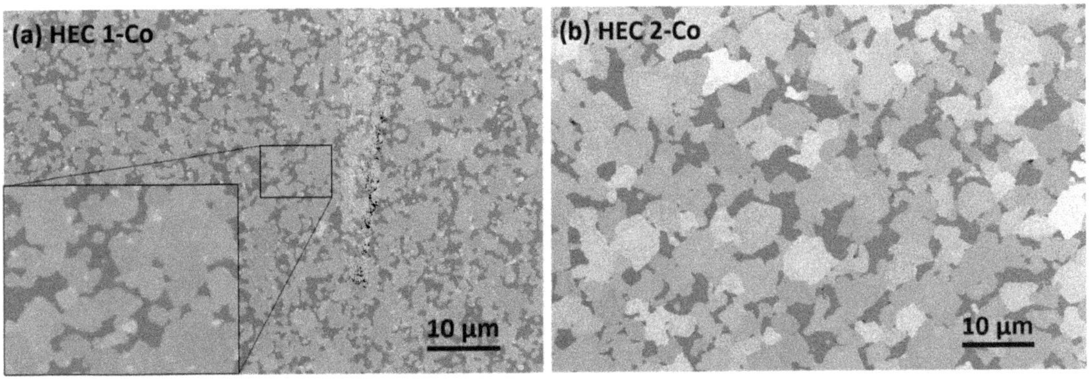

Figure 2. Micrographs of HEC-Co samples: (**a**) HEC 1-Co (composition: (Hf-Ta-Ti-Nb-V)C and Co) (**b**) HEC 2-Co (composition: (Ta-Ti-Nb-V-W)C and Co) both sintered with sinterHIP cycle at 1400 °C and 100 bar Ar pressure. (The dark gray phase is the cobalt binder, see also Figures 3–6).

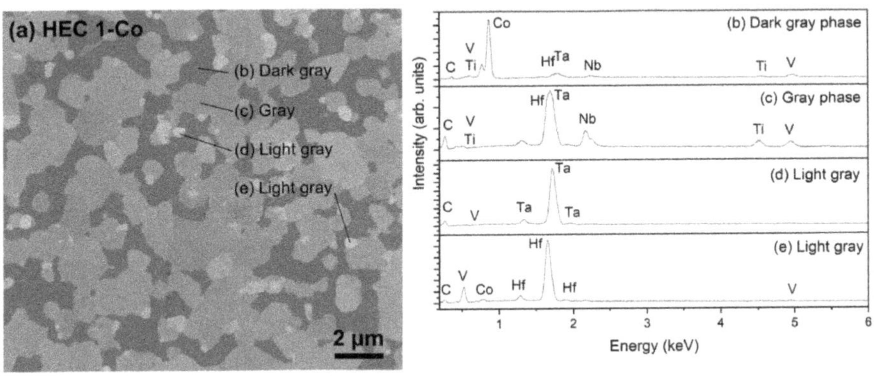

Figure 3. (**a**) Backscattered electron micrograph of HEC 1-Co with EDS mapping, (**b**) EDS result of dark phase consisting of Co with mixed metals, (**c**) EDS result of gray phase with mixed carbides, (**d**) EDS result of light gray phase with Ta and (**e**) EDS result of gray phase with Hf.

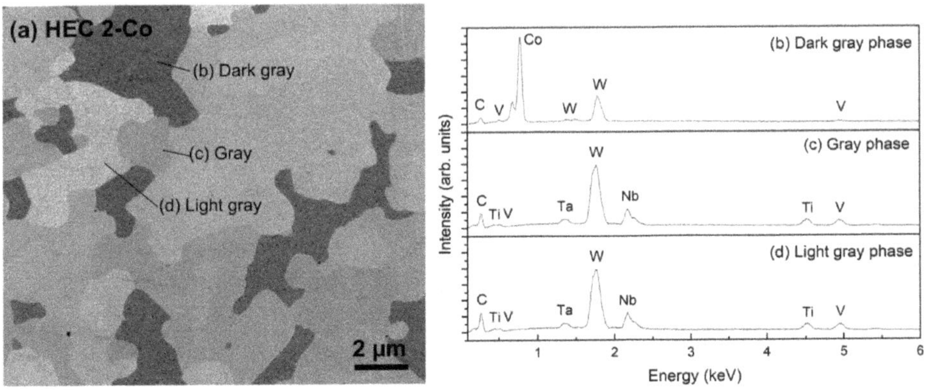

Figure 4. (**a**) Backscattered electron micrographs of HEC 2-Co with (**b**) EDS result of dark phase consisting of Co with mixed metals, (**c**) EDS result of gray phase with mixed carbides with high concentration of W, (**d**) EDS result of light gray phase same as (**c**).

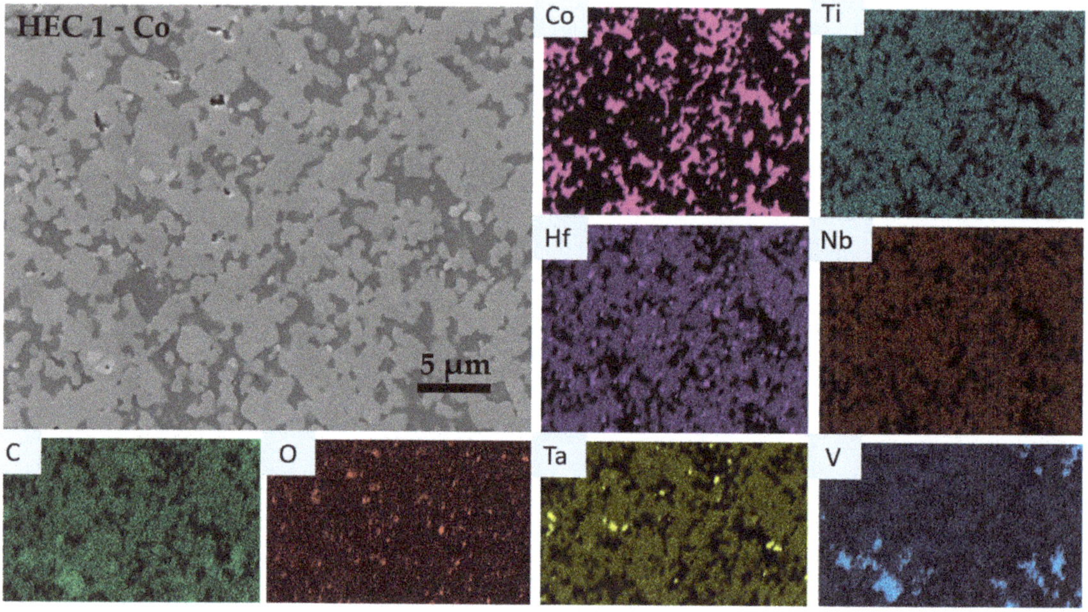

Figure 5. EDS micrographs of HEC 1-Co (composition: (Hf-Ta-Nb-Ti-V)C and Co, co-sintered with sinterHIP at 1400 °C and 100 bar Ar pressure).

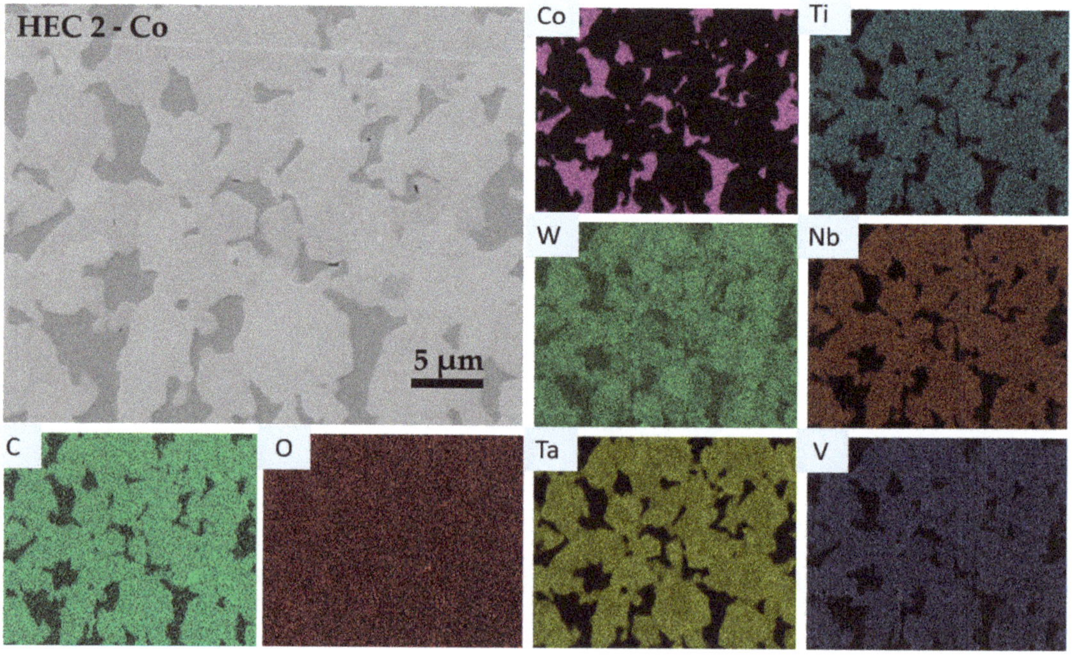

Figure 6. EDS micrographs of HEC 2-Co (composition: (Ta-Nb-Ti-V-W)C and Co, co-sintered with sinterHIP at 1400 °C and 100 bar Ar pressure).

Light gray spots could be observed at the grain boundaries of HEC 1-Co (Figure 2a). These could be carbide precipitations with a different composition. There are no light gray spots (as opposed to light gray phases) observed in HEC 2-Co (Figure 2b).

Figure 3 shows the SEM image of HEC 1-Co, the composition of which was (Hf-Ta-Ti-Nb-V)C with Co binder, and the corresponding EDS results. There are different gray levels shown in the electron micrographs. The respective element peaks in Figure 3b–d show that the dark gray phase is the cobalt binder with a low amount of dissolved carbide-forming metals (Hf, Ta, Nb, V and Ti), the gray phase is the complex solid solution of (Hf-Ta-Nb-Ti-V)C, and the light gray phase contains either Ta or Hf as the main component.

In most parts, the cubic carbides are all mixed together forming the hard phase, which is surrounded by Co (dark gray phase). The light gray round grains, which are rich in Hf and Ta, seem to be the remains of undissolved or just partially dissolved TaC and HfC from its corresponding HEC 1 powder.

The SEM images of HEC 2-Co (Figure 4a) show three kinds of phases: dark gray, gray, and light gray. EDS spectra of HEC 2-Co (Figure 4b–d) reveal that the dark gray phase is rich in Co and poor in other carbide-forming metals (Ta, Ti, Nb, V, W). The gray phase and light gray phase represent the high-entropy carbide grains, both rich in W. In this case, only an orientational contrast is responsible for the different gray shades.

To analyze the distribution of the elements in more detail, EDS mapping was carried out (Figures 5 and 6).

EDS mapping of HEC 1-Co (Figure 5) shows that next to the Co-based binder phase and the HEC phase, three other phases are present in minor quantities. A Ta carbide, as already found in HEC production, is most prominent, and is most likely the product of an insufficient homogenization during HEC production. Furthermore, V-Ti-rich carbide aggregates can be observed in some areas of the sample. Such areas were not spotted during the local EDS analysis of the bulk HEC samples. These aggregates could be the results of either a local precipitation during cooling, as found in V-containing WC-Co hardmetals [40,41], or a locally unfinished solid solution during HEC production. However, no V-Ti-based carbides were found in the prepared HEC powder or in the similarly produced binderless HEC ceramics [39]. The co-presence of Hf and O might point to a Hf-oxide phase. Both phases ((V,Ti) C and HfO_2) are in minor quantities, as proven by XRD analysis (Figure 7).

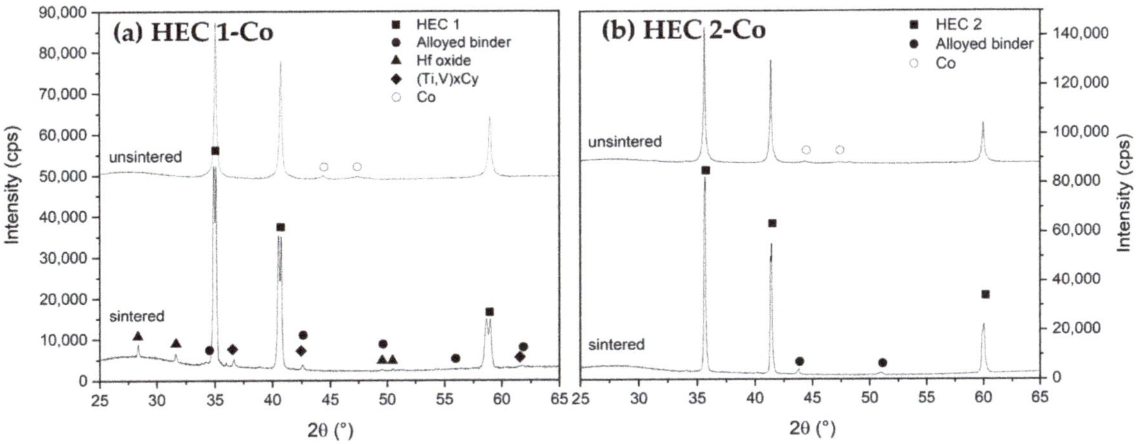

Figure 7. X-ray diffraction pattern of pressed (unsintered) and sintered (**a**) HEC 1-Co (composition: (Hf-Ta-Nb-Ti-V)C and 19.2 vol% Co) and (**b**) HEC 2-Co (composition: (Ta-Nb-Ti-V-W)C and 19.2 vol% Co) with 2Theta ranging between 25 and 65 degrees. Note: The labeled HEC 1 phase is in reality a mixture of two cubic carbide phases with very similar lattice parameters.

The HEC 2 starting powder consisted of (Ta-Ti-Nb-V-W)C and some excess carbo However, no excess carbon was detected in the micrographs of the sintered HEC 2-C (Figure 6). It is expected that the free carbon has been dissolved within the Co bind phase (possible C uptake up to ca. 4.1 at.% [42]) and has to some extent also reduced th surface oxides of the HEC and Co powders during sintering. All HEC starting element are homogenously distributed in the carbide phase. A slight variation in at least W and T can be detected in the color maps of these elements. This might correspond to the ligh color contrast variation in the carbide grains seen in the electron micrograph. It might ref to a solution precipitation mechanism and the beginning of the formation of a core–ri structure with a slightly W-depleted and slightly Ta-enriched rim area. No formation c separate carbide phases was observed.

Additional semiquantitative EDS-SEM analyses were carried out to obtain informatio about the stoichiometry of both HEC phases. In the case of HEC 1-Co, 30 HEC grains we measured. The ratios of the carbide-forming metal elements of the HEC 1 phase are give in Table 4. The data show no clear distinction between different compositions of the HE phase. However, some grains showed a variation in the vanadium content, resulting i the higher standard deviation. There is no distinct difference in the compositions of th different grains. The EDS mapping also revealed some V-rich grains. These grains we also analyzed (Table 4).

Table 4. Metal content (at.%) of HEC grains in HEC 1-Co determined by EDS-SEM.

Element	Ti	V	Nb	Hf	Ta
Main phase					
Mean value/at.%	20.2	18.2	21.5	19.3	20.9
Standard deviation	0.9	3.2	1.2	1.1	1.9
V- rich phase					
Mean value/at.%	14.7	66.9	9.6	1.3	7.5
Standard deviation	0.5	2.4	0.9	0.3	1.0

In the case of HEC 2-Co, the local formation of a core–rim structure was observe (Figure 6). Regions of the core and the rim were analyzed by measuring the carbide-formin metal elements of the HEC 2 phase of five core and rim areas each. The ratios are given i Table 5 and reveal some W and V depletion in the rim structure. In agreement with th EDS analysis of the binder phase (Figure 4), this suggests that the depletion is caused b the dissolution of W and V within the Co-binder phase.

Table 5. Metal content (at.%) of core and rim areas of HEC grains in HEC 2-Co determined b EDS-SEM.

Element	Ti	V	Nb	Ta	W
Core					
Mean value/at.%	18.3	19.5	21.2	20.4	20.7
Standard deviation	0.4	0.6	0.4	0.4	0.4
Rim					
Mean value/at.%	21.4	16.8	22.8	22.1	16.9
Standard deviation	1.1	0.8	0.7	1.0	2.0

Furthermore, the oxygen content was measured in all studied HEC grains of HE 1-Co and HEC 2-Co compositions by semiquantitative EDS-SEM analyses. This was done t investigate, if oxycarbides could have formed. The oxygen content for all 40 measurement varied from 0 to 2 at.% O. With the assumption that most of the measured oxygen cam from surface oxides formed after sample polishing, we consequently exclude the formatio of a significant incorporation of oxygen in the carbide lattice in our samples.

XRD measurements of the starting powder mixtures (as pressed) and of the sintered HEC 1-Co and HEC 2-Co samples are shown in Figure 7a,b. The diffraction pattern of the starting HEC 1-Co mixture (Figure 7a) shows three strong peaks, which correspond to the cubic HEC phase and the small peaks of the added Co phase. There are no matching PDF cards for the specific HEC composition. However, the lattice parameters are similar to the calculated lattice parameters using the rule of mixture (Vegards law) [39]. After sintering, the HEC phase is split up into two very similar phases, both having the cubic Fm-3m structure with very similar lattice parameters (0.4447 nm and 0.4422 nm), corresponding also to the lattice parameter of the HEC phase in the starting powder mixture. There are several other smaller peaks in the sintered HEC 1-Co sample. The peaks at 28.3° and 31.6° indicate the presence of monoclinic HfO_2 (PDF 01-075-6426), verifying the results of the EDS mapping in Figure 2. The EDS mapping of the HEC 1-Co sample also showed areas with a carbide enriched in V and Ti. Accordingly, a cubic Fm-3m (Ti,V)C matches the occurring peaks in the XRD pattern. The split-up of the HEC phase into two very similar HEC phases is most likely the result of the formation of the cubic Fm-3m (Ti,V)C phase and the subsequent loss of V and Ti in some HEC 1 grains, which were also observed in the mapping and analyzed by semiquantitative EDS-SEM.

However, this might not be the only reason for the splitting of the peaks of the main carbide phase. Therefore, we assume that the observed splitting could have also been caused by the formation of a core–rim structure which was not detected via SEM and EDS. It could also be caused by the different solubility of the components in Co. A slight adaptation of the starting composition thus might prevent the segregation. However, this needs further investigations.

The EDS analysis also showed that the cobalt binder is alloyed with metal atoms from the cubic carbides such as Ta and V, forming a cubic Co alloy with a changed lattice parameter in comparison to pure Co. This is confirmed by the XRD measurements, where a cubic phase is present with enlarged lattice parameters in comparison to pure Co.

In contrast to the sample HEC 1-Co, only a cubic HEC and a metallic Co-alloy binder were detected by XRD in the sample HEC 2-Co (Figure 7b). The cubic carbide solid solution has an Fm-3m structure. No splitting of the peaks of the HEC phase was observed. A slight HEC peak broadening was observed in the HEC 2-Co sample, most likely corresponding to the observed formation of the core–rim structure for some HEC 2 grains (Figure 6, Table 5). Similar to HEC 1-Co, the Co-binder is alloyed with metal atoms originating from the metal carbides, forming a cubic Co phase with slightly changed lattice parameters.

3.3. Physical and Mechanical Properties of HEC-Co Hardmetals

An ideal high-entropy hardmetal would be one wherein hard phase and binder phase are homogeneously distributed and balanced between hardness and toughness. The hardness and toughness can also be influenced by the carbon content of the sintered hardmetal, which can be measured by magnetic saturation (mS). In addition, the measurement of coercivity (Hc) determines the degree of sintering, cobalt distribution, and grain size of hardmetals. The values of mS, Hc, hardness, and fracture toughness for the produced HEC-Co hardmetals are given in Table 6 in comparison with a WC-12 wt. % Co hardmetal. The theoretical magnetic saturation (mS) was calculated based on the cobalt weight content of both high-entropy hardmetal compositions.

Table 6. Physical and mechanical properties of HEC-Co hardmetals.

Samples	mS $/\mu T \cdot m^3 \cdot kg^{-1}$	mS /(% theo)	Hc /kA/m	Hardness /HV10	Hardness GPa	Fracture Toughness $K_{1C}/MPa \cdot m^{1/2}$
HEC 1-19.2 vol% Co	30.0	70.0	5.6	1203 ± 15	11.80	8.5 ± 0.2
HEC 2-19.2 vol% Co	20.7	57.5	8.4	1432 ± 10	14.04	6.7 ± 0.2
WC-19.2 vol% Co [43]	19.9	85.0	19.9	1460 ± 15	14.32	14.3 ± 0.2

In Table 4, the relative mS for HEC 1-Co (70%) and HEC 2-Co (57.5%) represents a lower limit where the Eta phase would be present in the case of WC-Co hardmetals. However, since it is not clear which of the carbide-forming elements in the HEC hard phase dissolves in the binder phase, a clear assessment cannot be made without a larger variety of added carbon contents. Since no Eta phase of any form was detected by EDS or XRD, it can still be expected that both samples are within a two-phase region of HEC and Co in a corresponding theoretical HEC-Co-C diagram. In the case of a WC-Co hardmetal, the coercivity of ca. 6 and 8 kA/m would correspond to a medium to coarse hardmetal grade of 1.3 to 2.5 μm and 2.5 to 6.0 μm, respectively.

The HV10 hardness values for HEC 1-Co and HEC 2-Co are both lower than for a corresponding medium-grain-sized WC-12 Co hardmetal, due to the larger grain size of the produced materials. Additionally, the measured fracture toughness is somewhat lower than that of the corresponding WC-Co-based hardmetal.

4. Conclusions and Outlook

Due to the growing interest in high-entropy carbides, an investigation of HEC-based hardmetals was done for the first time. Two different high-entropy carbides with the composition (Hf-Ta-Ti-Nb-V)C and (Ta-Ti-Nb-V-W)C were synthesized by a gas pressure sintering technique (also called sinterHIP) at 1950 °C and 100 bar Ar pressure. These high-entropy carbides were crushed to particle sizes less than 20 μm, mixed with 19.2 vol% Co (equivalent to WC-12 wt. % Co), and sintered using a sinterHIP cycle at 1400 °C and 100 bar Ar pressure. The EDS mapping of the resulting dense HEC-based hardmetals with the composition (Hf-Ta-Ti-Nb-V)C-Co (HEC1-Co) showed a homogenous HEC-Co microstructure, as well as a slight amount of undissolved TaC, HfO_2, and some larger vanadium–titanium–carbide segregations (already present in the starting HEC).

For the HEC 1-Co material ((Hf-Ta-Ti-Nb-V)C-Co), XRD measurements reveal the formation of a secondary HEC phase with a very similar composition. The reason for the splitting is still not clear, but might be caused by the solution–precipitation process known from hardmetals and cermets, and the formation of a core–rim structure.

However, for the HEC 2-Co material ((Ta-Ti-Nb-V-W)C-Co), homogeneously mixed carbide grains embedded in a binder phase were observed. Only a broadening of the XRD peaks of the HEC phase indicates that the formation of a core–rim structure of the HEC phase started. This is in agreement with the SEM investigation, which revealed a slight depletion of some of the carbide-forming elements in the rim area. A slight adaptation of the starting composition might prevent this process.

The measurement of magnetic saturation indicates for both HEC hardmetals a larger dissolution of HEC-forming elements, with resulting mS values in the range of 60 to 70% as compared to unalloyed cobalt. Coercivity indicated a medium to coarse HEC grain size. The hardness value for HEC 1-Co ((Hf-Ta-Ti-Nb-V)C) is 1200 HV10, while the hardness value for HEC 2-Co ((Ta-Ti-Nb-V-W)C) is 1430 HV10. With up to 8.5 MPa·$m^{1/2}$, the fracture toughness values are lower than those of WC-Co hardmetals with corresponding hardness. However, with a further carbon adjustment or with other binder compositions such as Ni, Ni, Co, Ni, Mo or Ni, Fe higher fracture toughness values are expected, since Ni has been shown to enhance the mechanical properties in cermets made from cubic hard phases such as TiC, TiCN and NbC [4,44].

Further investigations are necessary to fully understand the microstructure–property relations in these materials. The slightly selective solution–precipitation of some components in certain HECs, for example, might open up possibilities to influence the binder properties and to improve the properties of the material.

In conclusion, dense HEC-based hardmetals have not only been successfully fabricated for the first time by sinterHIP, but for the first time at all. This shows that HEC-based hardmetals can be produced in industrial-style furnaces, which leads to a high number of new possibilities, such as material freedom for the manufacturing of machining tools and other kinds of tools. Future research will focus on the use of directly synthesized HEC powders, different HEC compositions and the use of other binder elements, including high-entropy alloys such as CoCrFeNiCu.

Author Contributions: Conceptualization, J.P. and M.D.; methodology, J.P., A.V. and M.H.; validation, J.P., A.V. and M.H.; formal analysis J.P. and A.V.; investigation, M.D., J.P. and A.V.; data curation, M.D., J.P. and A.V.; writing—original draft preparation, J.P., M.D. and A.V.; writing—review and editing, J.P. and M.H.; visualization, M.D., A.V. and J.P.; supervision, A.M. and M.H.; project administration, J.P. and M.H.; All authors have read and agreed to the published version of the manuscript.

Funding: This research received no external funding.

Institutional Review Board Statement: Not applicable.

Informed Consent Statement: Not applicable.

Acknowledgments: The authors thank Markus Mayer and Daniel Reichenbach as well as Björn Matthey for their help with sample preparations and analysis.

Conflicts of Interest: The authors declare no conflict of interest. The funders had no role in the design of the study; in the collection, analyses, or interpretation of data; in the writing of the manuscript, or in the decision to publish the results.

References

1. Antonov, M.; Hussainova, P. Chromium carbide based cermets as the wear resistant materials. In Proceedings of the 4th International Conference Industrial Engineeringinnovation as Competitive Edge for SME, Tallinn, Estonia, 29–30 April 2004; pp. 169–172.
2. Huang, S.G.; van der Biest, O.; Li, L.; Vleugels, J. Properties of NbC–Co cermets obtained by spark plasma sintering. *Mater. Lett.* **2007**, *61*, 574–577. [CrossRef]
3. Huang, S.G.; Vanmeensel, K.; Mohrbacher, H.; Woydt, M.; Vleugels, J. Microstructure and mechanical properties of NbC-matrix hardmetals with secondary carbide addition and different metal binders. *Int. J. Refract. Met. Hard Mater.* **2015**, *48*, 418–426. [CrossRef]
4. Huang, S.G.; Vleugels, J.; Mohrbacher, H.; Woydt, M. NbC grain growth control and mechanical properties of Ni bonded NbC cermets prepared by vacuum liquid phase sintering. *Int. J. Refract. Met. Hard Mater.* **2018**, *72*, 63–70. [CrossRef]
5. Montenegro, P.; Gomes, J.; Rego, R.; Borille, A. Potential of niobium carbide application as the hard phase in cutting tool substrate. *Int. J. Refract. Met. Hard Mater.* **2018**, *70*, 116–123. [CrossRef]
6. Pötschke, J.; Richter, V.; Mayer, M. Manufacturing and properties of polycrystalline WC-Co based cemented carbides. In Proceedings of the 2016 International Conference on Powder Metallurgy & Particulate Materials, MPIF, Boston, MA, USA, 5–8 June 2016; pp. 457–466.
7. Ettmayer, P. Hardmetals and Cermets. *Annu. Rev. Maler. Sci.* **1989**, *19*, 145–164. [CrossRef]
8. Hummer, J. WC-CrC-Co Sintered Composite. Patent No. US3480410A, 15 May 1968.
9. Hashe, N.G.; Norgren, S.M.; Andrén, H.-O.; Neethling, J.H. Characterization of WC–(W,V)C–Co made from pre-alloyed (W,V)C. *Int. J. Refract. Met. Hard Mater.* **2009**, *27*, 229–233. [CrossRef]
10. Holleck, H.; Thümmler, F. Eigenschaften von (Ta. Mo)- und (Ta. W)-Mischcarbiden und deren Verhalten zu Bindemetallen. *Plansee Semin.* **1975**, *23*, 175–185.
11. Kieffer, R.; Trabesinger, G.; Reiter, N. Über NbC- und HfC-haltige Hartstoffe und Hartmetalle. *Plansee Semin.* **1969**, *17*, 25–35.
12. Exner, H. Physical and chemical nature of cemented carbides. *Int. Met. Rev.* **1979**, *4*, 149–173.
13. Norgren, S.; García, J.; Blomquist, A.; Yin, L. Trends in the P/M hard metal industry. *Int. J. Refract. Met. Hard Mater.* **2015**, *48*, 31–45. [CrossRef]
14. Ettmayer, P.; Kolaska, H. Cermets der neuen Generation. *Metall* **1989**, *43*, 742–749.
15. Lengauer, W.; Chen, L.; Gracia, J.; Ucakar, V.; Drezer, K.; Kassel, D.; Daub, H.W. Diffusion controlled surface modification for fabrication of functional gradient (Ti W)C based cemented carbonitrides. *Euro Powder Metall.* **1999**, *3*, 475–482.
16. Komac, M.; Kos, K.; Novak, S. Hard Metals Based on Titanium and Molybdenum Carbides. *Inst. Mater. Festkörp.* **1983**, *15*, 76–85.
17. Ettmayer, P.; Kolaska, H.; Lengauer, W.; Dreyer, K. Ti(C,N) cermets—Metallurgy and properties. *Int. J. Refract. Met. Hard Mater.* **1995**, *13*, 343–351. [CrossRef]
18. Lengauer, W.; Binder, S.; Aigner, K.; Ettmayer, P.; Guillou, A.; Debuigne, J.; Groboth, G. Solid state properties of group IVb carbonitrides. *J. Alloy. Compd.* **1995**, *217*, 137–147. [CrossRef]
19. Agnew, S.R.; Dong, L.; Keene, J.I.; Wadley, H.N.G. Mechanical properties of large TiC-Mo-Ni cermet tiles. *Int. J. Refract. Met. Hard Mater.* **2018**, *75*, 238–247. [CrossRef]
20. Chen, X.; Xiong, W.; Qu, J.; Yang, Q.; Yao, Z.; Huang, Y. Microstructure and mechanical properties of (Ti,W,Ta)C-xMo-Ni cermets. *Int. J. Refract. Met. Hard Mater.* **2012**, *31*, 56–61. [CrossRef]

21. Csanádi, T.; Castle, E.; Reece, M.J.; Dusza, J. Strength enhancement and slip behaviour of high-entropy carbide grains durin micro-compression. *Sci. Rep.* **2019**, *9*, 10200. [CrossRef]
22. Andrén, H.-O. Microstructures of cemented carbides. *Mater. Des.* **2001**, *22*, 491–498. [CrossRef]
23. Holmström, E.; Lizárraga, R.; Linder, D.; Salmasi, A.; Wang, W.; Kaplan, B.; Mao, H.; Larsson, H.; Vitos, L. High entropy alloy Substituting for cobalt in cutting edge technology. *Appl. Mater. Today* **2018**, *12*, 322–329. [CrossRef]
24. De La Obra, A.G.; Avilés, M.A.; Torres, Y.; Chicardi, E.; Gotor, F.J. A new family of cermets: Chemically complex but microstru turally simple. *Int. J. Refract. Met. Hard Mater.* **2017**, *63*, 17–25. [CrossRef]
25. Yan, X.; Constantin, L.; Lu, Y.; Silvain, J.-F.; Nastasi, M.; Cui, B. (Hf 0.2 Zr 0.2 Ta 0.2 Nb 0.2 Ti 0.2)C high-entropy ceramics wi low thermal conductivity. *J. Am. Ceram. Soc.* **2018**, *101*, 4486–4491. [CrossRef]
26. Castle, E.; Csanádi, T.; Grasso, S.; Dusza, J.; Reece, M. Processing and Properties of High-Entropy Ultra-High Temperatu Carbides. *Sci. Rep.* **2018**, *8*, 8609. [CrossRef]
27. Harrington, T.J.; Gild, J.; Sarker, P.; Toher, C.; Rost, C.M.; Dippo, O.F.; McElfresh, C.; Kaufmann, K.; Marin, E.; Borowsk L.; et al. Phase stability and mechanical properties of novel high entropy transition metal carbides. *Acta Mater.* **2019**, *16* 271–280. [CrossRef]
28. Feng, L.; Fahrenholtz, W.G.; Hilmas, G.E. Low-temperature sintering of single-phase, high-entropy carbide ceramics. *J. Ar Ceram. Soc.* **2019**, *102*, 7217–7224. [CrossRef]
29. Lu, K.; Liu, J.-X.; Wei, X.-F.; Bao, W.; Wu, Y.; Li, F.; Xu, F.; Zhang, G.-J. Microstructures and mechanical properties of hig entropy (Ti0.2Zr0.2Hf0.2Nb0.2Ta0.2)C ceramics with the addition of SiC secondary phase. *J. Eur. Ceram. Soc.* **2020**, *4* 1839–1847. [CrossRef]
30. Wei, X.-F.; Liu, J.-X.; Li, F.; Qin, Y.; Liang, Y.-C.; Zhang, G.-J. High entropy carbide ceramics from different starting materia *J. Eur. Ceram. Soc.* **2019**, *39*, 2989–2994. [CrossRef]
31. Ye, B.; Wen, T.; Nguyen, M.C.; Hao, L.; Wang, C.-Z.; Chu, Y. First-principles study, fabrication and characterization c (Zr0.25Nb0.25Ti0.25V0.25)C high-entropy ceramics. *Acta Mater.* **2019**, *170*, 15–23. [CrossRef]
32. Bao, W.; Wang, X.-G.; Ding, H.; Lu, P.; Zhu, C.; Zhang, G.-J.; Xu, F. High-entropy M2AlC-MC (M=Ti, Zr, Hf, Nb, Ta) composit Synthesis and microstructures. *Scr. Mater.* **2020**, *183*, 33–38. [CrossRef]
33. Wang, K.; Chen, L.; Xu, C.; Zhang, W.; Liu, Z.; Wang, Y.; Ouyang, J.; Zhang, X.; Fu, Y.; Zhou, Y. Microstructure and mechanic properties of (TiZrNbTaMo)C high-entropy ceramic. *J. Mater. Sci. Technol.* **2020**, *39*, 99–105. [CrossRef]
34. Feng, L.; Fahrenholtz, W.G.; Hilmas, G.E.; Zhou, Y. Synthesis of single-phase high-entropy carbide powders. *Scr. Mater.* **2019**, *16* 90–93. [CrossRef]
35. Chicardi, E.; García-Garrido, C.; Gotor, F.J. Low temperature synthesis of an equiatomic (TiZrHfVNb)C5 high entropy carbide b a mechanically-induced carbon diffusion route. *Ceram. Int.* **2019**, *45*, 21858–21863. [CrossRef]
36. Chicardi, E.; García-Garrido, C.; Hernández-Saz, J.; Gotor, F.J. Synthesis of all equiatomic five-transition metals High Entropy Ca bides of the IVB (Ti, Zr, Hf) and VB (V, Nb, Ta) groups by a low temperature route. *Ceram. Int.* **2020**, *46*, 21421–21430. [CrossRe
37. García, J.; Collado Ciprés, V.; Blomqvist, A.; Kaplan, B. Cemented carbide microstructures: A review. *Int. J. Refract. Met. Har Mater.* **2019**, *80*, 40–68. [CrossRef]
38. Shetty, D.K.; Wright, I.G.; Mincer, P.N.; Clauer, A.H. Indentation fracture of WC-Co cermets. *J. Mater. Sci.* **1985**, *2* 1873–1882. [CrossRef]
39. Pötschke, J.; Dahal, M.; Herrmann, M.; Vornberger, A.; Matthey, B.; Michaelis, A. Preparation of High Entropy Carbides b Different Sintering Techniques. *J. Mater. Sci.* **2021**. under review.
40. Pötschke, J.; Mayer, M.; Richter, V. Formation of carbide segregations in nanoscaled hardmetals. In Proceedings of the Euro PN 2015: International Power Metallurgy Congress and Exhibition, Reims, France, 4–7 October 2015.
41. Toufar, C.; Schubert, W.D.; Hashiya, M.; Kubo, Y. On the Formation of Precipitations in Highly (Cr,V)-doped Cemented Carbide In Proceedings of the 18th International Plansee Seminar, Reutte, Austria, 3–7 June 2013; Sigl, L.S., Kestler, H., Wagner, J., Eds Plansee SE: Reutte, Austria, 2013.
42. Fernandez Guillermet, A. Thermodynamic Analysis of the Co-C System. In *Zeitschrift für Metallkunde*; Hanser: Munich, German 1987; Volume 78, pp. 700–709.
43. Vornberger, A.; Pötschke, J.; Gestrich, T.; Herrmann, M.; Michaelis, A. Influence of microstructure on hardness and therma conductivity of hardmetals. *Int. J. Refract. Met. Hard Mater.* **2020**, *88*, 105170. [CrossRef]
44. Woydt, M.; Huang, S.G.; Vleugels, J.; Mohrbacher, H.; Cannizza, E. Potentials of niobium carbide (NbC) as cutting tools and fo wear protection. *Int. J. Refract. Met. Hard Mater.* **2018**, *72*, 380–387. [CrossRef]

metals

MDPI

Article

Fatigue Behavior of $Zr_{58}Cu_{15.46}Ni_{12.74}Al_{10.34}Nb_{2.76}Y_{0.5}$ Bulk Metallic Glass Fabricated by Industrial-Grade Zirconium Raw Material

Shichao Zhou [1], Tao Zhang [2], Lugee Li [2], Jiedan Yang [3], Min Zhang [4], Chengyong Wang [5] and Yong Zhang [1,6,*]

1 The State Key Laboratory for Advanced Metals and Materials, University of Science and Technology Beijing, Beijing 100083, China; b20190484@xs.ustb.edu.cn
2 The New Material Research Institute, Dongguan Eontec Co., Ltd., Dongguan 523662, China; zhangtao@z-eon.com (T.Z.); lli@lli.cc (L.L.)
3 The Dongguan Eontec Co., Ltd., Dongguan 523662, China; dan@e-ande.com
4 Luoyang Advanced Manufacturing Industrial R&D Center Tianjin Research Institute for Advanced Equipment, Tsinghua University, Luoyang 471000, China; zhangmin@lamic.com.cn
5 Institute of Manufacturing Technology, Guangdong University of Technology, Guangzhou 510006, China; cywang@gdut.edu.cn
6 Shunde Graduate School of University of Science and Technology Beijing, Foshan 528399, China
* Correspondence: drzhangy@ustb.edu.cn; Tel.: +86-010-62333073

Abstract: In this work, the fatigue behavior of a low-cost $Zr_{58}Cu_{15.46}Ni_{12.74}Al_{10.34}Nb_{2.76}Y_{0.5}$ (at%) bulk metallic glass (BMG) fabricated by industrial-grade Zirconium raw material was investigated under three-point bending loading mode. X-ray, fatigue tests under different stress amplitude and fatigue fractography were conducted in order to characterize the amorphous structure, fatigue stress-life (S-N) curve and fracture mechanism, respectively. It is found that the X-ray diffraction (XRD) result showed a fully amorphous structure due to high glass-forming ability, cracks initiated from inclusions near the rectangular corners at tensile surfaces and the fatigue endurance limit (~168 MPa) and fatigue ratio (~0.13) termed as fatigue endurance limit divided by ultimate tensile strength in stress amplitude were comparable to the similar BMG prepared by high pure raw materials.

Keywords: bulk metallic glass (BMG); fatigue behavior; industrial-grade zirconium raw material

Citation: Zhou, S.; Zhang, T.; Li, L.; Yang, J.; Zhang, M.; Wang, C.; Zhang, Y. Fatigue Behavior of $Zr_{58}Cu_{15.46}Ni_{12.74}Al_{10.34}Nb_{2.76}Y_{0.5}$ Bulk Metallic Glass Fabricated by Industrial-Grade Zirconium Raw Material. *Metals* **2021**, *11*, 187. https://doi.org/10.3390/met11020187

Received: 31 December 2020
Accepted: 19 January 2021
Published: 21 January 2021

1. Introduction

First developed some 60 years ago, amorphous alloys, usually namely metallic glasses, have represented an interesting class of potential structural materials due to the lack of long-range orderly arrangement of atoms [1]. The unique microstructure has led to a range of intriguing properties, such as high strength, excellent hardness, large elastic elongation and other functional properties. Prior researchers had attempted to characterize the mechanical properties of thin ribbons and wires because of very high cooling rates necessary to prevent crystallization, until bulk metallic glasses (BMGs) were developed that could be fabricated at low cooling rates due to improved amorphous forming ability. Nevertheless, BMGs are not still widely applied in engineering fields up to date due to the major limitations of the corresponding service behavior, such as fatigue behavior.

As we all know, fatigue behavior for structural materials is very important and approximately 90 percent of all service failures associated with mechanical causes is fatigue [2]. However, compared with other mechanical properties of BMGs, the fatigue behavior studies are few. Starting from 1998, Gilbert et al. first reported the fatigue behavior of BMG and found that the fatigue endurance limit was low as ~8% of ultimate tensile strength [3]. From then, many researchers reported that the fatigue endurance limits of BMGs exhibited a wide range with 8–50%, usually less than 20% of their ultimate strength [2]. Such poor fatigue ratio was attributed to the lack of microstructure barriers such as grain boundaries and other lattice defects that can provide local crack–arrest positions [4–6]. Neverthe less, most of those were fabricated with high purity materials, let alone BMGs preparec with industrial raw materials. In fact, there are many factors that could be involved tc play roles in fatigue behavior for alloys, such as composition, material quality, specimer geometry, environment, temperature, surface condition, cyclic frequency, stress ratior and others [2]. Some of those might play more important roles than others. Hence, it is significant to clarify and understand the fatigue behavior of low-cost BMGs prepared witl industrial-grade materials.

Apart from that, low-cost raw materials will promote industrial production anc applications for BMGs. In the relevant studies about lowering the cost of BMGs, Jiang et al. [7–12] developed low-cost BMGs fabricated with industrial-grade raw materials by introducing a small number of rare earth elements to improve glass-forming ability. Based on this strategy, we designed and fabricated a low-cost $Zr_{58}Cu_{15.46}Ni_{12.74}Al_{10.34}Nb_{2.76}Y_{0.5}$ (at %) BMG with industrial-grade sponge Zirconium (Zr) and minor Yttrium (Y) additior characterizing remarkable amorphous forming ability and low-cost [13]. Accordingly, this study is focused on the fatigue behavior of a low-cost $Zr_{58}Cu_{15.46}Ni_{12.74}Al_{10.34}Nb_{2.76}Y_{0.5}$ (at %) BMG prepared by industrial-grade material, with the objective of characterizing the fatigue and fracture mechanisms.

2. Materials and Methods

In this study, the master alloys were firstly fabricated with industrial-grade sponge Zr by vacuum induction melting method. The elements of Cu (99.95%), industrial-grade sponge Zr (99.5%) with an oxygen content of less than 1000 ppm, Ni (99.96%), Al (99.9%) and Nb (99.95%) were adopted for the preparation of the ingots. To achieve a homogeneous distribution of elements, the master alloys were re-melted one time, flipped for each melt Prior to melting, the furnace chamber was evacuated and then backfilled with high purity Ar. The cooling in the melting was conducted in a water-cooled steel mold. After that, the alloys were produced by high-pressure die-casting and machined into rectangular fatigue test samples with a dimension of 2 mm × 2 mm × 30 mm. The schematic of samples and fatigue test loading conditions are shown in Figure 1. Additional details about the ingot preparation are described elsewhere [13].

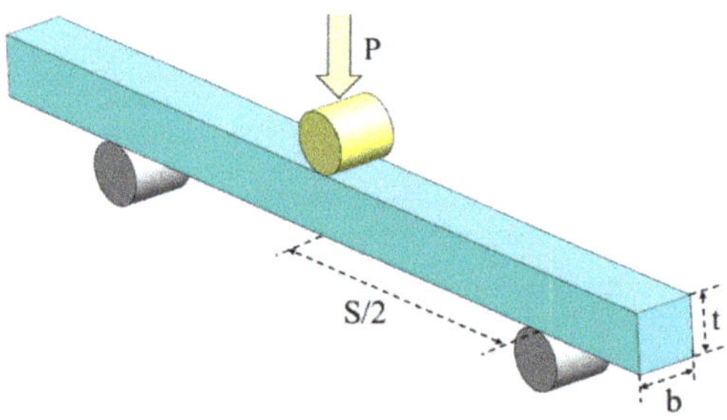

Figure 1. The schematic of samples and fatigue test mode.

Before fatigue testing, the specimens were ground with SiC papers and polished to produce mirror surfaces. Then, specimens were employed to conduct fatigue testing until the specimens failed or 10^7 cycles. Stresses were calculated from simple beam mechanics theory, using the following equation:

$$\sigma = \frac{3PS}{2bt^2} \quad (1)$$

where P, b, t and S are the applied load, width (~2 mm), thickness (~2 mm) and loading span (~20 mm) of fatigue test specimens, respectively. The fatigue tests were conducted by a computer-controlled material test system (MTS Acumen 3) electrohydraulic-testing machine (MTS Systems Corporation, Eden Prairie, MN, USA). In order to obtain the stress–life (S-N) fatigue data, specimens were tested under a wide range of maximum bending stress from 340 MPa to 1000 MPa (just below the ultimate tensile strength) with a constant R ratio (R = $\sigma_{min}/\sigma_{max}$, where σ_{max} and σ_{min} are the applied maximum and minimum stresses, respectively) of 0.1 under a stress-controlled mode in the air at ambient temperature with a sinusoidal waveform at a loading frequency of 60 Hz. To compare fatigue ratio with other alloys, tensile tests were conducted and the tensile stress–strain curve of this BMG at a constant strain rate of 1×10^{-3}/s under ambient temperature is shown in Figure 2.

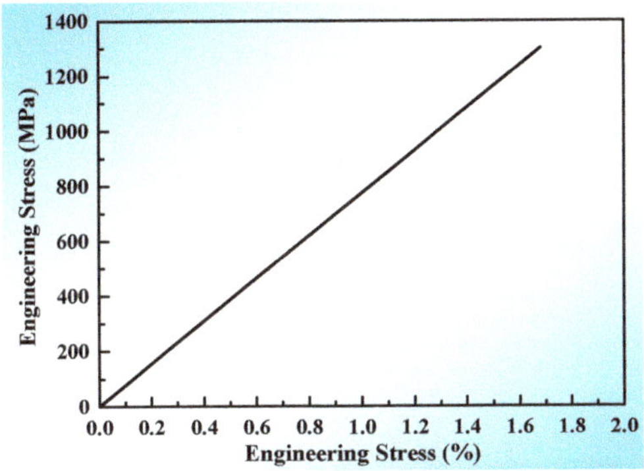

Figure 2. Tensile engineering stress–strain curve for $Zr_{58}Cu_{15.46}Ni_{12.74}Al_{10.34}Nb_{2.76}Y_{0.5}$ bulk metallic glass (BMG).

The amorphous structure was determined by standard X-ray diffraction (XRD, Rigaku, Tokyo, Japan) with a TTP III (Cu K_α radiation) in the range of 20–80°, with a step-size of 0.02°, a dwelling time of 0.5 s and scan speed 5°/min. The failure surfaces of specimens were observed by scanning electron microscopy (SEM, Carl Zeiss Microscopy Ltd, Jena, Germany) with a Zeiss Supra 55 operated at 20 keV and a working distance of 15 mm.

3. Results

3.1. Calculation for Parameters Ω and δ

Based on the phase formation rule for multi-component alloys proposed by Yang and Zhang [14], relevant ΔS_{mix}, Ω and δ parameters for our BMG were calculated to estimate the glass formation ability. With high ΔS_{mix} (≈1.2R), δ (≈10) and low Ω (≈0.5), this BMG coincides with BMG-formation criterion, indicating high glass-forming ability. Here, ΔS_{mix}, Ω and δ parameters are defined as below:

$$\Delta S_{mix} = -R\sum_{i=1}^{n} c_i \ln c_i \quad (2)$$

$$\Omega = T_m \Delta S_{mix}/|\Delta H_{mix}| \quad (3)$$

$$\Delta H_{mix} = \sum_{i=1, i\neq j}^{n} c_i c_j \Omega_{ij} \quad (4)$$

$$\Omega_{ij} = 4\Delta H_{AB}^{mix} \quad (5)$$

$$\delta = 100\sqrt{\sum_{i=1}^{n} c_i \left(1 - \frac{r_i}{\bar{r}}\right)^2} \quad (6)$$

where c_i is mole percent of component, R is gas constant (R = 8.314 $J \cdot K^{-1}$ mol^{-1}), T_m is th melting temperature of the multi-component alloys, ΔH_{AB}^{mix} is the enthalpy of mixing binary liquid alloys, r_i is atomic radius and $\bar{r}$ is the average atomic radius.

3.2. XRD Pattern

The XRD pattern (Figure 3) shows a fully amorphous structure for our BMG fabricate by industrial-grade sponge Zr, which is also consistent with phase criteria [14].

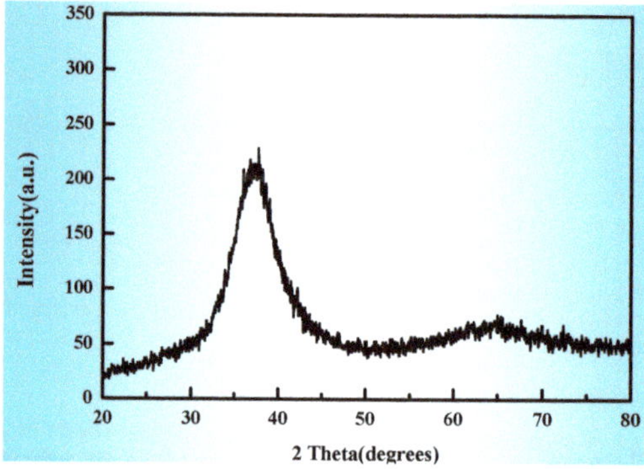

Figure 3. X-ray diffraction pattern of the $Zr_{58}Cu_{15.46}Ni_{12.74}Al_{10.34}$ $Nb_{2.76}Y_{0.5}$ BMG specimen.

3.3. Stress–Life (S-N) Fatigue Data

The stress–life (S-N) fatigue data of this BMG is shown in Figure 4. Figure 4a show S-N curve in the form of applied maximum stress, σ_{max}, versus the number of cycles t failure, N_f. In order to compare fatigue endurance limit with other alloys, as shown i Figure 4b, S-N curves present in terms of N_f versus fatigue ratio (σ_a/σ_u), stress amplitud σ_a (σ_a = 1/2 ($\sigma_{max-}\sigma_{min}$) normalized by the ultimate fracture strength (σ_u) of alloys, a shown in Equation (7), where σ_{max} and σ_{min} are the maximum and minimum values of th applied loading cycle [15]. It is worth noting that σ_u represents ultimate tensile strengt under three-point bending, four-point bending and tension–tension fatigue modes, an ultimate compressive strength under compression–compression fatigue mode, respectivel Hence, the fatigue endurance limits (fatigue ration) under different loading modes can b compared with σ_a/σ_u. As shown in Figure 4a, the normalized fatigue endurance limit c our BMG is σ_{max} = 373 MPa. The ultimate tensile strength of this BMG is ~1300 MPa a ambient temperature, as seen from Figure 2. As shown in Figure 4b, the normalized fatigu endurance limit of our BMG is $\sigma_a/\sigma_u \approx 0.13$ and $\sigma_a \approx 168$ MPa (black line). This is compa rable to Vitreloy 105 ($Zr_{52.5}Ti_5Cu_{17.9}Ni_{14.6}Al_{10}$, $\sigma_a/\sigma_u \approx 0.13$, blue line) with high purit raw materials under compression–compression mode [16] and $Zr_{52.1}Ti_5Cu_{17.9}Ni_{14.6}Al_{10}Y_{0.}$ BMG (our previous work, $\sigma_a/\sigma_u \approx 0.14$, green line) with industrial-grade raw Zr mate rial under three-point bending mode [17]. However, it is apparent that all of them ar lower than Vitreloy 105 with high purity raw materials under four-point bending mod ($\sigma_a/\sigma_u \approx 0.24$, red line) [18]. Apart from BMGs, commercial aluminum alloy (2020-T81, ye low line) and steel (300-M, pink line) were compared [3]. However, the fatigue enduranc limit of our BMG was lower than those. Moreover, the values of σ_{max} and the cycles t failure of our BMGs were fitted for engineering applications, as shown in Equation (8).

$$\sigma_a/\sigma_u = \frac{1}{2}\frac{\sigma_{max} - \sigma_{min}}{\sigma_u} \quad (7$$

$$\log(N_f) = 7.0861 - 1.1465\log(\sigma_{max} - 372.1) \quad (8$$

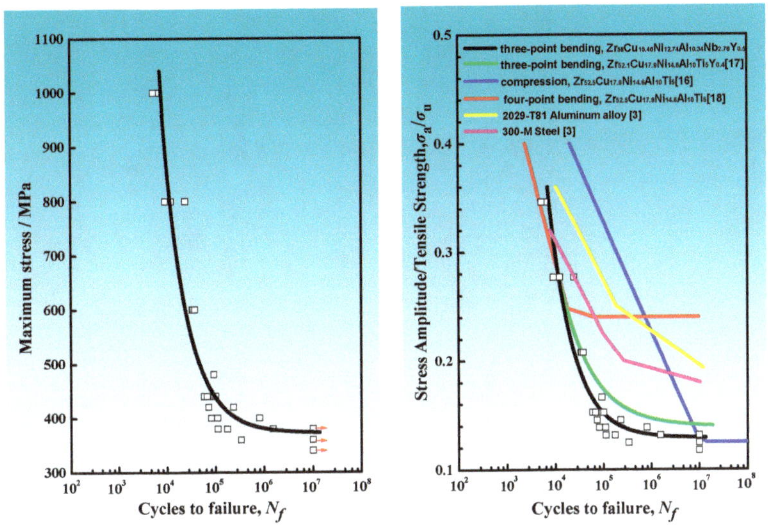

(**a**) (**b**)

Figure 4. Stress–life (S-N) fatigue data for several alloys. (**a**) Maximum stress (MPa) versus N_f; (**b**) Stress amplitude/tensile strength (σ_a/σ_u).

3.4. Fatigue Fractography

Figure 5 shows the fatigue fractography of our BMG. The fatigue fractography tested at σ_{max} = 800 MPa after N_f = 24,369 are shown in Figure 5a–c and σ_{max} = 400 MPa after N_f = 805,992 are shown in Figure 5d–f, respectively. Both of the fracture surfaces are perpendicular to the tensile stress direction. It is apparent that the fractured surface shows three typical regions: crack initiation region, stable crack propagation region, and final unstable fast fracture region. Figure 5a,d show over fatigue fractography and magnification images of crack initiation region; Figure 5b,c,e,f show magnification images of crack propagation region and fast fracture region, respectively. As shown in Figure 5a,d, the fatigue crack initiated from the tensile side surface (outer side) of specimens and the initiation sites were near the corner of beams. Further detailed observation showed some as-cast defects, inclusions, near the crack initiation sites. In the latter stage, the fatigue crack propagated and then formed an elliptical crack, as shown in Figure 5a,d. The area of this region after long cycles life (N_f = 805,992, as shown in Figure 5d) covered larger than short cycles life (N_f = 24,369, as shown in Figure 5a). In this stable crack propagation region, typical uniform fatigue striations also were observed, as shown in Figure 5b,e. In the final stage, in the fast fracture region, typical vein-like patterns were observed as shown in Figure 5c,f.

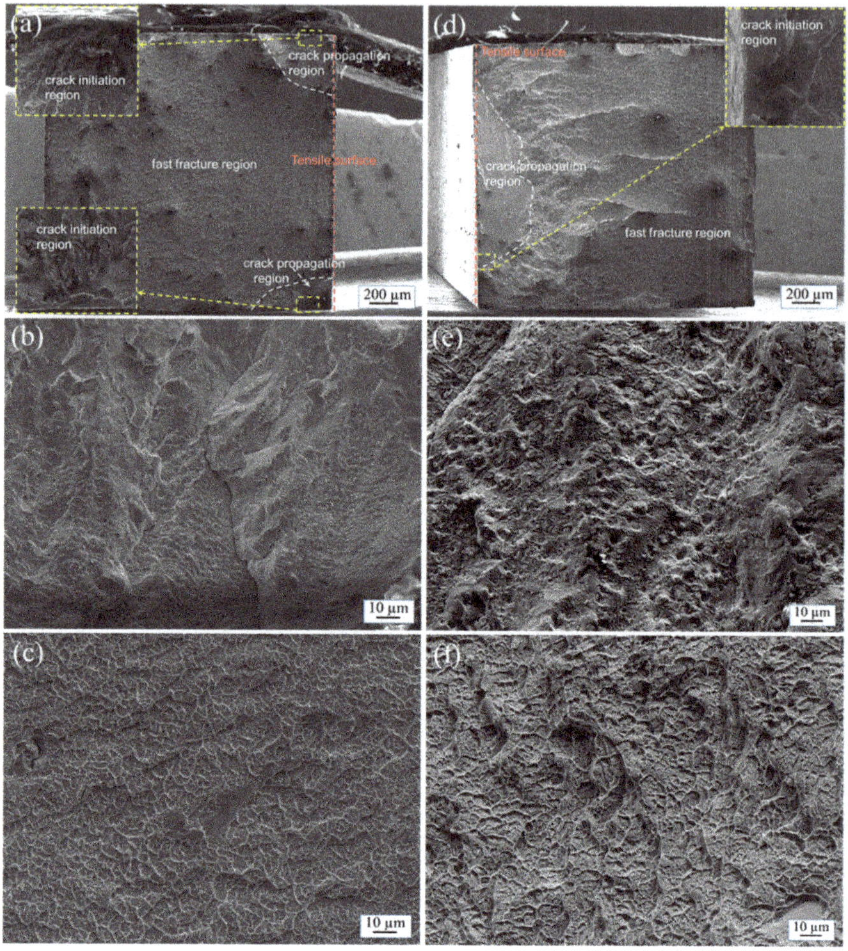

Figure 5. Fatigue fractography of our bulk metallic glass (BMG) tested at σ_{max} = 800 MPa after N_f = 24,369 as shown in (**a–c**) and σ_{max} = 400 MPa after N_f = 805,992 as shown in (**d–f**). (**a**,**d**) show over fatigue fractography and magnification images of crack initiation region; (**b**,**e**,**c**,**f**) show magnification images of crack propagation region and fast fracture region, respectively.

4. Discussion

Several factors should be considered in the differences of fatigue endurance limits for BMGs, such as composition, material quality, sample geometry, experiment environment, surface condition, cyclic frequency, ultimate fracture strength and others. Zr_{58} $Cu_{15.46}$ $Ni_{12.74}$ $Al_{10.34}$ $Nb_{2.76}$ $Y_{0.5}$ BMG (this work) and $Zr_{52.1}Ti_5Cu_{17.9}Ni_{14.6}Al_{10}Y_{0.4}$ BMG (our previous work) possessed similar fatigue endurance limits due to similar compositions and ultimate tensile strength. Generally speaking, fatigue endurance limits are positively associated with ultimate tensile/compressive strength. Thus, the two BMGs possess similar S-N curves. In addition, as shown in Figure 4b, the fatigue lifetimes under compression–compression mode (though possessed the lowest fatigue endurance limit) are longer than that under three-point bending and four-point bending mode. According to other research [16], loading mode may be the dominant reason. The fatigue crack initiation and propagation can be impeded under compression–compression mode. It is possible that the fatigue crack propagated under compression–compression mode like a Mode-II crack, but Mode-I crack propagated under bending mode [19]. Mode-I crack is an opening or tensile mode tending to propagate quickly. However, Mode-II crack tends to slid or shear mode and propagates relatively slowly. Therefore, the lifetimes may be diverse due to different loading modes for BMGs. The fatigue endurance limit of Vitreloy 105 BMG under four-point bending mode (red line) is higher than others, as shown in Figure 4b. The major reason for the great variation in fatigue endurance limits under similar loading modes may be specimen geometry and defects [17,19]. Although there are all beam samples, the corners of the samples are different. Rounded corners under four-point bending mode can strongly reduce the stress concentration as opposed to rectangular under three-point bending and compression–compression modes. It is possible that the fatigue cracks in the rectangular samples formed more easily than those in the rounded samples. Then, the inclusions, as shown in Figure 5a,d, played an important role and further enhance stress concentration. The free volume and incompatible deformation could be caused near the interface between inclusions and matrix under cyclic loading due to the composition diversity. Subsequently, the stress concentration was further enhanced, which promoted the formation of cracks. There are two main views about fatigue crack initiation. Cracks may initiate from the shear bands [2,20–23] or casting defects, such as inclusions and pores. It is apparent that, in our low-cost BMG with industrial-grade Zr raw material, fatigue crack initiated from inclusions due to stress concentration. Typical uniform fatigue striations were obvious on the stable crack propagation region, as shown in Figure 5b,e. For crystalline alloys, the striation formation is associated with the blunting and resharpening of the crack tip during each cyclic loading. Similarly, the striations in BMGs are also associated with the blunting and resharpening process [18]. The vein-like patterns were observed in the unstable fast fracture region, as seen in Figure 5c,f. The abundant vein-like pattern formed in this region is due to the viscosity change suddenly. At the moment of the failure, abundant elastic energy could be released, meanwhile, the temperature rises sharply [24]. Thence, the viscosity greatly reduced and vein-like patterns formed. However, there are no distinct local melting regions observed in other research studies [25,26]. It is possible that the released elastic energy may not provide enough heat to result in local melting.

5. Conclusions

In this study, the fatigue behavior of a low-cost $Zr_{58}Cu_{15.46}Ni_{12.74}Al_{10.34}Nb_{2.76}Y_{0.5}$ (at%) BMG fabricated by industrial-grade Zr raw material was investigated under three-point bending loading mode. According to the relevant results and discussion above, the following conclusions can be drawn.

With high ΔS_{mix} ($\approx$1.2 R), δ ($\approx$10) and low Ω ($\approx$0.5), $Zr_{58}Cu_{15.46}Ni_{12.74}Al_{10.34}Nb_{2.76}Y_{0.5}$ BMG possesses high glass-forming ability. The fracture surface shows three typical regions: crack initiation region containing inclusions, stable crack propagation region consisting of striations, and final unstable fast fracture region characterizing vein-like pattern. The crack initiated from inclusions near the rectangular corners due to stress concentration. The fatigue endurance limit (~168 MPa) in stress amplitude and fatigue ratio (~0.13) of this BMG are comparable to the similar BMG (Vitreloy 105) prepared by high pure raw materials.

Author Contributions: S.Z., writing—original draft, data analysis, investigation and methodology; T.Z., preparation for specimens and data analysis; L.L., data analysis; J.Y., investigation; M.Z., methodology and investigation; C.W., editing; Y.Z., conceptualization, supervision. All authors have read and agreed to the published version of the manuscript.

Funding: This work was financially supported by Guangdong Basic and Applied Basic Research Foundation (2019B1515120020); National Natural Science Foundation of China (51735003).

Institutional Review Board Statement: Not applicable.

Informed Consent Statement: Informed consent was obtained from all subjects involved in the study.

Data Availability Statement: The data presented in this study are available on request from th corresponding author.

Conflicts of Interest: The authors declare no conflict of interest.

References

1. Jun, W.; Willens, R.; Duwez, P. Non-crystalline structure in solidified gold-silicon alloys. *Nature* **1960**, *187*, 869–870. [CrossRef]
2. Jia, H.; Wang, G.; Chen, S.; Gao, Y.; Li, W.; Liaw, P. Fatigue and fracture behavior of bulk metallic glasses and their composite *Prog. Mater. Sci.* **2018**, *98*, 168–248. [CrossRef]
3. Gilbert, C.; Lippmann, J.; Ritchie, R. Fatigue of a Zr-Ti-Cu-Ni-Be bulk amorphous metal: Stress/life and crack-growth behavi *Scr. Mater.* **1998**, *38*, 537–542. [CrossRef]
4. Gilbert, C.; Schroeder, V.; Ritchie, R. Mechanisms for fracture and fatigue-crack propagation in a bulk metallic glass. *Metall. Mat Trans. A* **1999**, *30*, 1739–1753. [CrossRef]
5. Guennec, B.; Nobori, T.; Kuwahara, H.; Ueno, K. Effect of the stress ratio on the fatigue behavior of $Zr_{55}Al_{10}Ni_5Cu_{30}$ bulk metall glass part I—Analysis of the fatigue resistance. *Intermetallics* **2018**, *92*, 72. [CrossRef]
6. Menzel, B.; Dauskardt, R. Stress-life fatigue behavior of a Zr-based bulk metallic glass. *Acta Mater.* **2006**, *54*, 935–943. [CrossRe
7. Jiang, F.; Wang, Z.; Zhang, Z.; Sun, J. Formation of Zr-based bulk metallic glasses from low purity materials by scandium additio *Scr. Mater.* **2005**, *53*, 487–491. [CrossRef]
8. Zhang, Y.; Pan, M.; Zhao, D.; Wang, R.; Wang, W. Formation of Zr-based bulk metallic glasses from low purity of materials b yttrium addition. *Mater. Trans.* **2000**, *11*, 1410–1414. [CrossRef]
9. Zhang, Y.; Zhou, Y.; Hui, X.; Wang, M.; Chen, G. Minor alloying behavior in bulk metallic glasses and high-entropy alloys. *S China. Ser. G* **2008**, *4*, 427–437. [CrossRef]
10. Zhang, Y.; Chen, J.; Chen, G.; Liu, X. Glass formation mechanism of minor yttrium addition in CuZrAl alloys. *Appl. Phys. Le* **2006**, *89*, 131904. [CrossRef]
11. Zhang, Y.; Li, R. Editorial for special issue on nanostructured high-entropy materials. *Int. J. Miner. Metall. Mater.* **2020**, 2 1309–1311. [CrossRef]
12. Yang, M.; Liu, X.; Wu, Y.; Wang, H.; Wang, X.; Lu, Z. Unusual relation between glass-forming ability and thermal stability c high-entropy bulk metallic glasses. *Mater. Res. Lett.* **2018**, *6*, 495–500. [CrossRef]
13. Zhang, T.; Meng, X.; Wang, C.; Li, L.; Yang, J.; Li, W.; Li, R.; Zhang, Y. Investigations of new bulk metallic glass alloy fabricated using a high-pressure die-casting method based on industrial grade Zr raw material. *J. Alloys Compd.* **2019**, *79* 851–859. [CrossRef]
14. Yang, X.; Zhang, Y. Prediction of high-entropy stabilized solid-solution in multi-component alloys. *Mater. Chem. Phys.* **2012**, *13* 233–238. [CrossRef]
15. Launey, M.; Hofmann, D.; Johnson, W.; Ritchie, R. Solution to the problem of the poor cyclic fatigue resistance of bulk metall glasses. *Proc. Natl. Acad. Sci. USA* **2009**, *106*, 4986–4991. [CrossRef]
16. Wang, X.; Qu, R.; Wu, J.; Liu, Z.; Zhang, Z. Fatigue damage and fracture behavior of metallic glass under cyclic compressio *Mater. Sci. Eng. A Struct.* **2018**, *717*, 41–47. [CrossRef]
17. Zhou, S.; Zhang, T.; Zhang, M.; Zhang, Y. Fatigue behavior of a minor yttrium doped ZrCuNi-based metallic glass alloy fabricate by industrial grade raw material. *MRS Adv.* **2020**, *5*, 1713–1721. [CrossRef]
18. Naleway, S.; Greene, R.; Gludovatz, B.; Dave, N.; Ritchie, R.; Kruzic, J. A highly fatigue-resistant Zr-based bulk metallic glas *Metall. Mater. Trans. A* **2013**, *44*, 5688–5693. [CrossRef]
19. Wang, G.; Qiao, D.; Yokoyama, Y.; Freels, M.; Inoue, A.; Liaw, P. Effects of loading modes on the fatigue behavior of Zr-base bulk-metallic glasses. *J. Alloys Compd.* **2009**, *483*, 143–145. [CrossRef]
20. Wang, X.; Qu, R.; Liu, Z.; Zhang, Z. Shear band-mediated fatigue cracking mechanism of metallic glass at high stress level. *Mate Sci. Eng. A Struct.* **2015**, *627*, 336–339. [CrossRef]
21. Wang, X.; Qu, R.; Liu, Z.; Zhang, Z. Evolution of shear-band cracking in metallic glass under cyclic compression. *Mater. Sci. En A Struct.* **2017**, *696*, 267–272. [CrossRef]
22. Qu, R.; Wang, S.; Wang, X.; Wang, Z.; Zhang, Z. Revealing the shear band cracking mechanism in metallic glass by X-ra tomography. *Scr. Mater.* **2017**, *133*, 24–28. [CrossRef]
23. Wang, X.; Qu, R.; Wu, S.; Zhu, Z.; Zhang, H.; Zhang, Z. Improving fatigue property of metallic glass by tailoring the microstructur to suppress shear band formation. *Materialia* **2019**, *7*, 100407. [CrossRef]
24. Wang, G.; Liaw, P.; Yokoyama, Y.; Freels, M.; Inoue, A. Investigations of the factors that affected fatigue behavior of Zr-base bulk-metallic glasses. *Adv. Eng. Mater.* **2008**, *11*, 1030–1033. [CrossRef]
25. Qiao, D.; Fan, G.; Liaw, P.; Choo, H. Fatigue behaviors of the $Cu_{47.5}Zr_{47.5}Al_5$ bulk-metallic glass (BMG) and $Cu_{47.5}Zr_{38}Hf_{9.5}Al$ BMG composite. *Int. J. Fatigue* **2007**, *29*, 2149–2154. [CrossRef]
26. Zhang, L.; Chen, Z.; Chen, D.; Zhao, X.; Zheng, Q. Four-point-bending-fatigue behavior of the $Cu_{45}Zr_{45}Ag_7Al_3$ bulk metalli glass. *J. Non-Cryst. Solids* **2013**, *370*, 31–36. [CrossRef]

MDPI
St. Alban-Anlage 66
4052 Basel
Switzerland
Tel. +41 61 683 77 34
Fax +41 61 302 89 18
www.mdpi.com

Metals Editorial Office
E-mail: metals@mdpi.com
www.mdpi.com/journal/metals

www.ingramcontent.com/pod-product-compliance
Lightning Source LLC
Chambersburg PA
CBHW062102220926
51850CB00010B/2857
* 9 7 8 3 0 3 6 5 2 4 0 6 1 *